AF388668

Assessing the Performance of Passive Houses in Mediterranean Climate Regions

Assessing the Performance of Passive Houses in Mediterranean Climate Regions

Special Issue Editor

Vincenzo Costanzo

MDPI • Basel • Beijing • Wuhan • Barcelona • Belgrade • Manchester • Tokyo • Cluj • Tianjin

Special Issue Editor
Vincenzo Costanzo
Struttura Didattica Speciale di Architettura— Università degli Studi di Catania
Italy

Editorial Office
MDPI
St. Alban-Anlage 66
4052 Basel, Switzerland

This is a reprint of articles from the Special Issue published online in the open access journal *Sustainability* (ISSN 2071-1050) (available at: https://www.mdpi.com/journal/sustainability/special_issues/passive_houses).

For citation purposes, cite each article independently as indicated on the article page online and as indicated below:

LastName, A.A.; LastName, B.B.; LastName, C.C. Article Title. *Journal Name* **Year**, *Article Number*, Page Range.

ISBN 978-3-03928-949-3 (Hbk)
ISBN 978-3-03928-950-9 (PDF)

Cover image courtesy of Vincenzo Costanzo.

Contents

About the Special Issue Editor

Vincenzo Costanzo, Ph.D. MEng, is currently Researcher at the Department of Electric, Electronic and Computer Engineering (DIEEI) at the University of Catania (Italy), where he was also awarded his PhD in Energetics. He is author of several journal papers published on international peer-reviewed journals and of many contributions on national and international conference proceedings in the fields of building energy efficiency, building performance simulation, thermal and visual comfort, renewable energy production in buildings, and daylighting. He has been involved in numerous international research projects, including the "REELCOOP" (Renewable Electricity Cooperation, FP7 funded) and "LoHCool" (Low Heating and Cooling of Buildings, EPSRC/NSFC funded) projects in his previous appointment as Research Associate at the University of Reading (UK).

Preface to "Assessing the Performance of Passive Houses in Mediterranean Climate Regions"

The Passive House (PH) concept was born in Germany in the last decade of the 20th century, with the aim of providing better indoor air quality (IAQ) and thermal comfort conditions than average constructions but at a minimum expense of operational energy.

Building upon traditional bioclimatic, low-energy, and solar building design approaches—which suggest the judicious use of design techniques and materials to heat, cool, and light the building with a reduced or null amount of energy (i.e., a passive design approach)—the PH Institute codified a set of performance targets required for a building to reach the Passive House standard.

Several studies reported on the performance achieved by PH buildings in operation, generally showing a good agreement between design expectations and experimental measurements. Significant deviations in the design predictions have been mainly attributed to incorrect realizations (e.g., not perfectly airtight windows and the presence of thermal bridges) and occupant behavior (e.g., non-optimal operation of setpoints, opening of windows and shading operation).

However, deviations from design overheating occurrences have also been widely reported during the winter season. The overheating issue is even more important when considering PH realizations in hotter climates, despite the PH standard generically advising on the possibility to reduce the amount of insulation of the envelope, raise non-adventitious air infiltration, and adopt additional typical passive design solutions.

This book presents a series of original research studies from authors around the world, where the issues in applying the PH Standard for the design of buildings in warm Mediterranean climate are tackled. Case studies as well as numerical analyses are presented and thoroughly discussed to stimulate ideas for building scientists and designers to achieve their design performance goals.

Vincenzo Costanzo
Special Issue Editor

Article

The Passivhaus Standard in the Spanish Mediterranean: Evaluation of a House's Thermal Behaviour of Enclosures and Airtightness

Víctor Echarri-Iribarren [1,*], Cristina Sotos-Solano [1], Almudena Espinosa-Fernández [2] and Raúl Prado-Govea [1]

[1] Department of Building Construction, University of Alicante, Carretera de San Vicente s/n, San Vicente del Raspeig, 03690 Alicante, Spain
[2] Department of Architecture, University of Zaragoza, Calle María de Luna, 3, 50018 Zaragoza, Spain
*   Correspondence: Victor.Echarri@ua.es; Tel.: +34-965-903677

Received: 3 June 2019; Accepted: 4 July 2019; Published: 8 July 2019

Abstract: Few houses have been built in the Spanish Mediterranean in accordance with the Passivhaus (PH) standard. This standard is adapted to the continental climates of Central Europe and thorough studies are necessary to apply this standard in Spain, especially in the summer. High relative air humidity levels in coastal areas and solar radiation levels of west-facing façades require adapted architectural designs, as well as greater control of air renewal and dehumidification. A priori, energy consumptions undergo big variations. In this study, the construction of a single-family house in the Spanish Levante was analysed. All enclosure layers were monitored using sensors of surface temperature, solar radiation, indoor and outdoor air temperature, relative humidity, and air speed. The thermal behaviour of the façade enclosure and air infiltration through the enclosure were examined using the blower door test and impacts on annual energy demand were quantified. Using simulation tools, improvements are proposed, and the results are compared with examples of PH housing in other geographical areas. The annual energy demand of PH housing was 69.19% below the usual value for buildings in the Mediterranean region. Very thick thermal insulation and low values of airtightness could be applied to the envelope, which would work very well in the winter. These technique solutions could provide optimal comfort conditions with a well-designed air conditioning system in summer and low energy consumption.

Keywords: Passivhaus; thermal transmittance of enclosures; air infiltration; annual energy demand; energy efficiency

---

1. Introduction

Within the European Union, the energy consumption of buildings represents 40% of final energy consumption. Buildings are responsible for 36% of CO_2 emissions. A total of 85% of the energy consumed is used to heat or cool rooms, illuminate them, and heat sanitary water. In December 2002, the European Parliament approved the Energy Performance Building Directive, EPBD 2002/91/EC, with the aim of improving energy efficiency [1]. Long-term strategies have been established to decrease energy consumption and to reduce CO_2 emissions, based on the Kyoto Protocol. Both exterior and local weather conditions are taken into account, as well as internal climatic conditions and the cost-effectiveness ratio. A few years later, a revised text was approved, Directive 2010/31/EU, known as the EPDB Recast [2]. Thus, the regulations on Nearly Zero Energy Buildings, known as NZEBs, were introduced for all new buildings from year 2020 onwards [3]. This text was revised in 2012 and has now become Directive 2012/27/EU [4].

All European Union member states made the commitment to submit themselves to the same directive and the 2020 goal is shared. However, the way of implementing the regulatory framework is defined by each country individually. Each country also individually establishes the criteria to assess buildings' energy efficiency and the needs of buildings with almost zero energy consumption. As the specific way of applying the EPDB is thus dispersed, not all countries situate themselves within the same ranges of energy consumption and CO_2 emissions, which is added to the fact that the energy efficiency evaluation criteria also differ.

Spain, as a member of the European Union, is today undergoing an 'energy transition'. The country is currently implanting renewable energy and energy efficiency in the national electricity system. The ultimate objective is to reshape the energy sector so that renewable energies account for 20% of final gross energy consumption in 2020 and 27% by 2030, with a possible increase of 30%. Spain's legislative framework regulating the energy efficiency of buildings consists in the Technical Building Code (or CTE by its Spanish acronym) [5]. This code came into effect in 2006 and was modified in 2009 [6], 2010, and more substantially in 2013 [7], to comply with the European Union's successive guidelines. A major change introduced in 2013 is the reduction of the maximum thermal transmittance U value allowed in a building envelope's elements. Heating and cooling energy consumption has also been restricted. In addition, types of climate according to the building's location and user profiles have been defined, both for residential and non-residential buildings. Currently, a new modification is underway whereby the energy efficiency strategies of the EU's building guidelines will be approved for nearly zero energy consumption buildings.

1.1. The Passivhaus Standard

A variety of standards define the requirements that buildings must comply with to guarantee minimum energy efficiency values. The most widespread is the Passivhaus (PH) standard. It was developed in Central European countries and the European Union uses the requirements found within it as design principles for energy efficient buildings [8].

The key PH standard requirements are as follows:

1. Annual heating demand must not exceed 15 kWh/m^2;
2. Annual demand for primary energy for heating, electricity and DHW must be less than or equal to 120 Kwh/m^2;
3. The volume of filtered air should not be greater than 0.6 air changes per hour (ACH), measured with a pressure of 50 Pa, as verified by an on-site test using the Blower Door test.

Further requirements were added to apply the standard to warmer climates, as follows:

4. Annual cooling demand must not exceed 15 kWh/m^2;
5. Thermal comfort must be achieved in all rooms of the house throughout the year. Therefore, the regulation allows 10% of indoor air temperature overheating, representing 10% of the total hours of a complete one-year cycle above 25 °C.

The main problem when adapting buildings on the Mediterranean coast to the PH standard is that in northern and central European countries, energy demands focus exclusively on heating. In southern Europe, however, air conditioning or cooling needs must be taken into account in summer, in addition to heating demands in winter. It is also worth noting that a large part of the population lives on the coast, with high relative humidity values, and buildings require high dehumidification values of outdoor renewal air, with its corresponding increase in energy demand in summer [9]. For this reason, a large amount of research and development projects focus on adapting homes with passive systems in warmer climatic conditions [10], in which these techniques are combined with PH standard requirements. Some of the most relevant projects are as follows:

CEPHEUS project: Study of more than 100 homes in five northern European countries [11]. Low levels of energy consumption were obtained, while occupants' thermal comfort remained optimal [12].

Passive-On project [13]: It focused on implementing the PH concept in southern European climates. The passive house concept was found to be viable, provided essential prior studies are conducted to adapt and detail the technical and constructive solutions in each specific region [14].

MED-ENEC: Different procedures were implemented through demonstration projects to show how to solve construction sector challenges in the southern Mediterranean region [15].

IEEA project: It addresses the passive strategies that could be adopted in different climatic zones of the EU [16].

ECO-BÂT project: Promoted by local authorities in Algeria [17]. It focuses on low-cost passive strategies and on affordable active measures that use renewable energy technologies.

Southern European passive houses: It was found that a pleasant indoor climate could be achieved by preconditioning the ventilation system's supply air [18]. Night ventilation has the same effect as façade insulation, especially when applying a mechanical ventilation system equipped with a by-pass.

Due to its successful implementation in Central Europe, the PH standard has been applied in other geographical and climatic zones, such as the Spanish Mediterranean region. In this sense, housing design under the Mediterranean climate must take into account parameters such as building orientation, solar gains in winter, thermal insulation of the envelope with a high thermal capacity, windows with double or triple glazing, solar control, control of thermal bridges to minimise them, both fixed and mobile solar protections, cantilevers, ventilation and natural lighting, etc. In addition, air renewal installations with double flow heat recovery systems must be improved.

The state-of-the-art regarding PH standard implementation provides interesting results that help determine its applicability in climates such as that of the Mediterranean coast. Figueiredo et al., after analysing a single-family, two-storey, 148 m^2 detached house of light construction in the region of Aveiro (Portugal), concluded that the problem of overheating of indoor air and walls constituted the main obstacle to achieving the PH standard in Portugal [19]. They emphasised the need to use a shading and solar protection system, in addition to increasing the enclosures' thermal inertia, thus attenuating thermal wave oscillation of the interior air temperature T_i. Costanzo, Fabbri, and Piraccini studied a house in Cesena (Italy) and concluded that its design, drawn up using the Passive Housing Planning Package (PHPP), was able to guarantee suitable comfort conditions during the heating period [20]. However, on the other hand, overheating during the cooling season was recorded for almost 50% of the time, according to the EN 15251 standard [21], an excessively high value. This value could be reduced to 20% by reducing the insulating material thickness (up to one third of the original value) on the roof and walls, replacing triple glazed windows with double glazed windows, and implementing a hybrid ventilation strategy instead of using mechanical ventilation with heat recovery (MVHR). Suárez, Prieto, and Salgado studied a house on the northern coast of Spain [22]. If the house were to be renovated, installing a ventilation system with heat recovery, improving the insulation of opaque building elements, replacing existing windows with high thermal efficiency windows, and reducing air infiltration rates to below 0.6 ACH, the annual energy demand would be below that required by the PH standard and the heating and cooling needs would be respectively reduced by 81% and 57% of the current values. With regard to the coast of Catalonia (Spain), Pesic, Roset, and Muros proposed the methodology "Climate potential for natural ventilation" (CPNV). They achieved notable cooling energy savings by applying a mixed mode (or hybrid mode) and night ventilation techniques [23], which could be used to achieve, along with other construction techniques in the installations, the PH standard. This same line was adopted by Fokaides et al. in Cyprus, when they studied a house in Nicosia with 110 m^2 of usable floor area. The nocturnal natural ventilation significantly reduced summer period overheating, making it possible to comply with the EN 15251 standard. Users must be sufficiently trained to adapt their behaviour to the home's thermal needs [24]. The application of the standard in Algeria throws some light on the issue. Ali-Toudert and Weidhaus concluded that night and day ventilation of 4 ACH was needed. The energy demand in winter was low, with a maximum value of 5.5 kWh/m^{2}yr, while in summer it was higher, with 9.7 kWh/m^{2}yr [25]. Other studies in climates with less hot summers focus on the use of passive and active systems [26], such as Mihai et al. in Bucharest, with earth-to-air heat exchange technology (EAHX) with photovoltaic panels, generating 1,556.5 kWh/year, and a drop in annual energy demand to 13 kWh/m^{2}yr [27]. Harkouss, Fardoun, and Biwole, through multi-criteria decision making (MCDM), further refined the technical quantification of thermal insulation and thermal transmittance U values in warm climates at 0.6 W/m^{2}K for walls, 0.6 W/m^{2}K for ceilings, and 0.5 W/m^{2}K for floors [28]. In mixed climates, the walls and ceiling should be well insulated (U= 0.2 W/m^{2}K), while the floor's thermal transmittance should not be minimised so as to allow the heat to escape through the floor in summer. A low window/surface area ratio value (10%) is necessary to improve buildings' energy performance.

From this review of the state-of-the-art, we can conclude that, to date, few PH dwellings have been built on the Spanish Mediterranean coast. Some of the research has focused on applying bioclimatic techniques in order to comply with the PH standard and to improve climatic conditions in summer, while others have limited themselves to quantifying energy demand in summer and winter. However, the absence of more detailed quantifications, according to the different parameters and intervening agents, of the energy consumption produced in these PH dwellings is detected. The novelty of this research is the quantification of energy consumption due to the transmission of heat through the various elements of the enclosures and of the actual infiltration of air through the enclosure by means of the analysis of a representative PH case study.

1.2. Aims and Objectives

The present study examined the implementation of the Passivhaus standard in the Spanish Mediterranean. It focused on the impact that building envelope heat flows and air infiltration have on annual energy demand [29]. To this end, the construction of an isolated detached house in Molina de Segura (Spain), located 40 km away from Spain's Eastern coast was studied. The house was designed with the aim of complying with the Passivhaus standard. The façade enclosures were monitored and thermal transmittance measurements were made in situ. The Blower Door air infiltration test was also conducted. Subsequently, using the Design Builder simulation tool [30], results on the quantification of the impact on energy demand were obtained for both parameters, both in the studied house and in the hypothetical case that it would have been built in the city of Alicante, on the coast [18]. Since the standard was designed to improve energy efficiency under a continental climate, the intention was to draw a conclusion on whether the standard could also be appropriate for coastal climate situations. Relative air humidity in the months of summer and autumn is much higher than that experienced in Central European countries, as well as wall and indoor air overheating by solar gains through glazing [31]. The thermal gaps between indoor air temperature and outdoor air temperature during the winter and spring months are more moderate. These big differences call into question, a priori, the validity of initial investments in the construction quality and the control devices for air renewal installations with heat recovery.

In addition to the tests mentioned previously, we proceeded to evaluate the parameters selected in this study in other houses in the area, according to local construction culture. First, we examined houses built before 1979 (without thermal insulation), then buildings built after the NBE CT-79 standard came into force [32], and finally, buildings constructed after 2006, after the CTE was implemented [5]. The enclosures of the latter buildings are resolved using two sheets with an air chamber and 6 cm of thermal insulation, an inverted flat roof with 6 cm of thermal insulation finished with gravel, and air permeable class 2 joineries (27 m^3/hm^2), with silicone-based sealing and double hollow brick walls, and high air infiltration standards according to the n_{50} index. In this way, results obtained in the present study can be compared with other construction systems and installations, allowing further studies to conclude on the suitability of the Passivhaus (PH) standard for Spanish Mediterranean coast buildings.

2. Description of the Case Study

The isolated detached house under study is located in Molina de Segura (Spain), 40 km from the coast (Figure 1). It is composed of two volumes. The lower volume has day areas (living room, laundry room, pantry, toilet, and office) and a floor area of 100 m^2. The first floor corresponds to the night area (3 bedrooms, 2 bathrooms, and a study) and has a constructed surface area of 95 m^2. There are covered porches on the outside with a constructed area of 43 m^2, which help to protect the house from solar radiation, resulting in a total floor area of 216.15 m^2. The bedrooms are south-facing to capture the largest possible amount of solar radiation in winter, while the living room and kitchen face a north-south axis, obtaining an axis of natural ventilation that passes through the house. Overheating in summer can thus be avoided (Figure 2). The gaps in the exterior façade have been planned in such a way as to capture a maximum amount of natural light. Overhangs and solar protections based on adjustable blades were designed to avoid direct solar radiation in summer. Flat roofs with small punctual overhangs protect the gaps from the impact of sunlight in summer.

Figure 1. View of the southeast façade. Main access to the house.

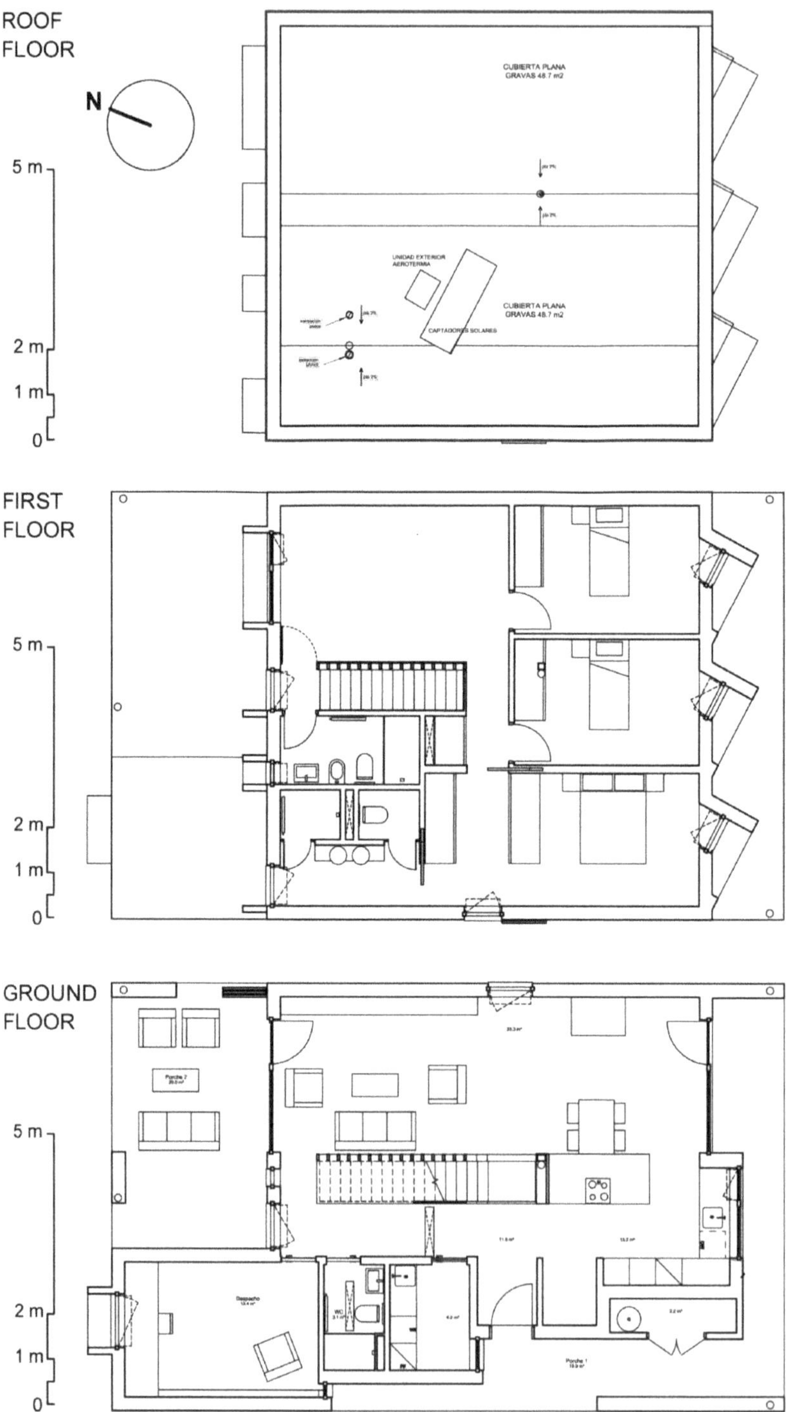

Figure 2. Floors of the house.

The façade on the ground floor is resolved with a thermal insulation system on the outside (SATE), thus constituting a ventilated façade made with a stucco lime mortar coating with flexible silicate on vertically modulated 32 × 100 mm mullions, a waterproof EPDM sheet, 15 mm Superpan Tech board, on a structure of modulated wooden posts every 60 cm and filled with natural 160 mm mineral wool, Micro USB barrier 230/20 Riwega, and laminated wood board on a self-supporting 73 × 48 structure (Figure 3). On the first floor, the exterior enclosure is a ventilated façade formed by larch wood slats, a 32 × 100 mm ventilation strip, Riwega waterproof sheet, 15 mm Superpan Tech board on a structure of modulated wooden mullions every 60 cm and filled with 148 mm natural mineral wool, and Micro USB 230/20 Riwega vapour barrier, with laminated wood board on a 73 × 48 self-supporting structure (Figure 4). All joineries between elements are solved with an adhesive tape seal, Riwega USB Tape 1 PE (60mm wide, professional and universal polyethylene with mesh reinforcement coated with high adhesive strength acrylic adhesive), which guarantees perfect air-tightness (Figure 5), as later verified with the results of the Blower Door test. The joinery is made of wood with thermal bridge break, elastic joint, Guardian-Sun type laminated glass (6 + 16 + 3 + 3), and Guardian-Sun type glass (6 + 16 + 4) for the non-laminated glazing.

Figure 3. View of the southeast façade. Construction of enclosures and floors.

The roof is flat, not passable, and is made with slopes, a SuperPan Tech P5 21 mm board, 200 gr/m^2 geotextile separating layer plus EPDM membrane waterproofing 1 mm thick, and a finishing layer of 5 cm-thick clean gravel, on a 198 × 48 wooden beam structure, 2 SuperPan Tech P4 25 mm wooden boards containing a 200 mm natural mineral wool thermal insulation layer, and a continuous suspended ceiling with a metal structure, formed by a laminated plaster plate. The foundation is based on a reinforced concrete slab, with a thermal insulation of XPS, 100 mm thick, and an EPDM waterproofing sheet, as well as self-levelling mortar and a class 3 microcement pavement.

Figure 4. Wooden structure and Superpan Tech 15 mm board partitions with 148 mm mineral wool thermal insulation.

The house has a low-temperature solar collector system for domestic hot water demand (DHW), in addition to auxiliary aerothermal equipment [33], located in the laundry room, which intervenes when there is insufficient solar radiation collection. In addition to natural ventilation, mechanical ventilation was installed with heat recovery with a contribution of 5 l/s per person in the bedrooms, 2 l/s per person in the living and dining rooms, and 2 l/s per m^2 in the kitchen, as specified in the CTE-DB-HS3. One air conditioning system is included, coupled to the ventilation system, and composed of an exterior machine as well as an indoor one, with a direct expansion cold-only air-air system. The EER value is 5.93 and the COP is 4.27.

Figure 5. Lacquered aluminium carpentry joined to the exterior pavement. Airproof sealing and waterproofing with EPDM (left). Interior partitions and enclosure with tape sealing, type Riwega USB Tape 1 PE (right).

2.1. Climatic and Normative Conditions

The house location is under a Mediterranean climate. According to the Köpen–Geiger classification [34], it is classified as BSh, a warm semi-arid climate with hot or very hot summers and mild winters with very little rainfall [35]. The average annual temperature is 18.1 °C, the coldest month being January, with an average of 10.3 °C, and the hottest month being August, with an average of 26.3 °C.

According to the Spanish regulations in force at the time of building the CTE house (2013 modification), it is located in a B3 climatic zone with an altitude of 240 m above sea level, letter B in winter and 3 in summer, according to the CTE-DB-HE table B1 classification (Climate Zones).

2.2. Energy Demands Obtained in the Project Phase

According to the current calculation procedure in Spain, CTE DB-HE-2013, using the Unified Tool Lider-Calener (HULC), the demand for renewable energy consumption was 29.7 kWh/m^2·year. The annual heating demand was 6.41 kWh/m^2·year, below the demand level required by the CTE (26.29 kW·h/m^2·year). Cooling energy demand was 4.17 kWh/m^2·year, i.e., below that required by the CTE (15 kWh/m^2·year).

3. Materials and Methods

The two domains of evaluation of the house's energy demand are linked to the materials used as well as the technical characteristics of the enclosures, although they are of a very different nature. Evaluating the actual heat flow through the enclosures throughout the year's entire cycle is a complex matter [36]. This evaluation requires simulation models that take into account construction faults, the impact of thermal inertia, real thermal bridges, as well as the incidence of solar radiation on the exterior surfaces and the air conditioning systems used. Air infiltration through the enclosures varies according to the external environment's climatic conditions, the air pressure at each moment, as well as how occupants maintain and use certain mobile elements such as joinery, adjustable slats, or grids. All this means that estimating the energy impact due to air infiltration is complex.

3.1. U Thermal Transmittance of the Enclosures

Fourier's law, which applies to parallel layers of materials, of an unlimited surface, in steady state indoor T_i and exterior T_e air temperatures, establishes the temperature gradient reached by multi-layer enclosures when there is a thermal gap between outdoor and indoor air temperatures (Equation (1)). Thermal transmission by conduction and convection intervene according to values of resistance to the passage of heat or superficial thermal resistance due to the convection currents generated.

$$U - value = \frac{1}{R_T} = \frac{1}{\frac{1}{h_i} + \sum_0^n \frac{e_i}{\lambda_i} + \frac{1}{h_e}}. \tag{1}$$

However, this law is valid only for ideal situations and steady states. Walls have limited surfaces, they present discontinuities due to joineries and glazing, thermal bridges due to construction faults [37,38], etc. In addition, the various layers are made up of different types of materials with different values of physical parameters of resistance to the passage of the heat [39]. Furthermore, outdoor air conditions are altered by solar radiation, air speed, water vapour pressure in the air, etc. Solar radiation also often has an impact on the outer surfaces, substantially increasing their temperature. Indoors, thermal loads and the air conditioning systems used alter the physical parameters of interior air temperature and relative humidity, as well as the surface temperatures of the walls [40]. Finally, the thermal inertia of the materials making up the enclosure alters the linear Fourier process as a function of time, since the thermal gap between exterior and interior air temperature, the solar radiation that affects the exterior surface, or the interior air conditioning systems vary according to the weather. Heat flow thus becomes a dynamic and complex process [41].

3.1.1. Theoretical Calculation of the U Transmittances of the Enclosures

In this study, we measured the house's thermal transmittance U value, but we did not analyse the effect of the parameters affecting the dynamic regime of the heat flow through the enclosures, such as thermal admittance, Y, and thermal wave lag [42], df, or the damping factor of the thermal wave, f_a. The thermal transmittance U values were obtained according to the method established by the current regulations, the Technical Building Code (CTE) [43].

The usual values of thermal conductivity, λ, of the materials used in the enclosure, according to UNE EN ISO 10456: 2012 [44,45], were compared to the measurements obtained from real samples by means of the thermal conductivity analyser C-Therm TCi by Mathis Instrumentos Ltd., using a universal sensor, made at the University of Alicante. We proceeded with a theoretical calculation of the heat flow resistance of the enclosure's different layers, under ideal conditions, according to Fourier's law.

To simulate this behaviour in a steady state, we overlooked the impact of other energy sources, such as solar radiation, the indoor conditioning system, or the effect of material effusiveness or thermal inertia [46], and decided to apply the maximum outdoor air temperature in summer (39.57 °C) and the minimum outdoor temperature in winter (7.87 °C), collected by PT 100 monitoring sensors, and an indoor air temperature of 28.62 °C in summer and 17.74 °C in winter, without the air conditioning system operating. To do this, we considered, in accordance with the CTE [47], that surface thermal resistance was *1/h_i* of 0.13 m^2 K/W, surface thermal resistance was *1/h_e* of 0.04 m^2 K/W, and thermal resistance, R_c, of the ventilated façade chamber was 0.0176 m^2 K/W.

3.1.2. In-Situ Measurements of U Transmittance

Subsequently, a thermoflumetric study of the opaque enclosures was conducted to measure U thermal transmittance [48,49]. It was measured in the two rooms on the southeast façade (the kitchen at a height of 0.70 m and the dining room), in the northwest façade (the living room), as well as in the bedrooms on the first floor in the northeast and southwest façades. For this, the ISO 9869-1: 2014 [50] standard was followed. The equipment consisted of a heat flow plate; a transducer that generates an electrical signal proportional to the total heat rate applied to the sensor surface. Two indoor and outdoor air temperature probes were attached to this plate, as well as two surface temperature sensors for the exterior and interior surfaces. These were connected to the analyser directly with cables or through the window. The trial lasted one week for each of the three measurements. The data was analysed using the module that calculates thermal transmittance, part of the AMR WinControl software developed by Ahlborn for ALMEMO measuring equipment. The method used was the "average method", which assumes that thermal transmittance or thermal conductivity can be calculated by dividing the average density of the thermal flow by the average temperature difference [51].

This analysis was completed with the thermal imaging and detection of possible thermal bridges, helping to determine the outer envelope's ventilated chamber behaviour [52,53], though quantifying thermal bridges was not the subject of this study [54].

3.2. Air Infiltration Through the Envelope

The air infiltration value through the enclosure was forecast to be very low thanks to the construction's quality, the seals used, and the joinery quality. It was evaluated by means of the Blower Door test conducted in accordance with the European Standard EN 13829, using the Blower Door GMBH MessSysteme für Luftdichtheit equipment. The corresponding graphs were obtained and there was an n_{50} air infiltration value of 50 Pa. To characterise the information obtained, the Minneapolis Blower Door Model 3 software (TEC-TITE 5.0) was used. This tool was developed specifically to describe 140 parameters that possibly intervene in the filtration phenomenon [55]. These parameters were collected in a database used to perform a general evaluation of the results.

To obtain the average value of air infiltration in the building, a parameter difficult to quantify, and to be able to perform the energy impact simulations using the Design Builder tool, thus obtaining the values of annual energy demand, the protocol established in the UNE-EN-ISO 13790: 2003 standard was followed, which it allows for converting the value of n_{50} to renewals/hour under normal pressure conditions by means of Equation (2):

$$n_{winter} = 2 \cdot n_{50} \cdot e_i \cdot \varepsilon_i \tag{2}$$

where:

n_{50} renewals/hour at 50 pascals;
e_i wind protection coefficient = 0.05; and
ε_i height corrector factor = 1.

The Blower Door test was the tool used to detect thermal bridges through images using a ThermaCam P 25 thermographic camera from Flyr. It was quantified using the AnTherm program. Load losses or gains due to thermal bridges were estimated at 3.5% of the total thermal loads by U thermal transmittance of the enclosures. These values are similar to values obtained in previous studies [56,57].

3.3. Energy Impact Assessment

As mentioned, the main objective of the present study was to quantify the energetic impact of the heat flow through the envelope and air infiltration. Both values were obtained using different tools and methodologies.

3.3.1. Energy Losses due to U Transmittance Values

To quantify the energetic impact, the transmittance U values of all the enclosures obtained in-situ were introduced into the Design Builder tool. To perform building behaviour simulations in terms of energy demand and interior comfort parameters, the data and parameter values that follow were introduced into the Design Builder tool. The winter period covered 1 December to 30 April and the summer period covered from 1 May to 30 November. This decision was based on results from previous studies, contrasted with the experience of various calibrations of models in the same geographical location. The interior air setpoint temperatures were 21 °C in winter and 24 °C in summer. For the standard calculation of air renewal, occupancy was 5 people. To comply with regulations, a value of 0.63 air changes per hour (ACH) of air renewal was established, in accordance with the Technical Building Code (CTE) currently in force. The air infiltration values introduced in the tool were those obtained from the Blower Door test. Table 1 shows the parameters introduced in Design Builder's simulation model.

Table 1. Summary of parameters introduced in Design Builder's simulation model.

Parameter	Data Introduced
Hours of operation and occupancy: Monday to Friday: 7 am to 3 pm	25 % occupancy
Hours of operation and occupancy: Monday to Friday: 3 pm to 8 pm	50 % occupancy
Hours of operation and occupancy: Monday to Friday: 8 pm to 7 am	100 % occupancy
Hours of operation, activity and occupancy: Saturday and Sunday	100 % occupancy
Climate equipment operating hours (heating)	7 am to 10 pm
Climate equipment operating hours (cooling)	9 am to 8 pm
Running of the air conditioning system from Monday to Sunday	7 days/week
Summer period	1 May to 30 November
Winter period	1 December to 30 April
Occupancy density (5 people)	0.03 person/m^2
Metabolic factor: "Standing/walking" option	1.20
Clothing values (CLO)	Winter CLO = 1.00 Summer CLO = 0.50
Load due to general lighting	300 lux
Internal air temperature setpoint Ti (cooling)	24 °C
Set internal air temperature Ti (heating)	21 °C
Indoor air maintenance temperature Ti (cooling)	26 °C
Internal air maintenance temperature Ti (heating)	18 °C
Relative humidity of the indoor air	50%
Air renewal rate	0.63 acH
Air infiltration through the envelope	0,049 acH

3.3.2. Annual Energy Demand due to Air Infiltration

Estimating the air infiltration's energy impact is a complex issue. It depends not only on the sealing of the building envelope, but also on weather conditions, which are sometimes very difficult to predict. No common criterion exists regarding the right model to evaluate the energetic impact of infiltrations. Until now, different calculation models have been developed with varying degrees of complexity and reliability. The simplest models presuppose a uniform distribution of leak paths and average constant leaks over time. In the present study, this impact was evaluated using a simplified model (Equation (3)), applying the concept of degree-day, which relates the average temperature outside the house under study and the interior comfort temperature (21 °C for heating and 24 °C for cooling) [7]. This estimate is theoretical and actual energy consumption depends on the particular conditions of the houses' internal air temperature setpoint, T_i. This calculation procedure allows evaluating the energetic impact, taking into account the climatic data specific to the area around the house. The air infiltration flow, specific air capacity, and the air temperature difference between the interior and the exterior of the house were all taken into consideration [58].

$$Q_{inf} = C_p \cdot G_t \cdot V_{inf}, \tag{3}$$

where:

Q_{inf} is annual energy loss (kWh/yr) due to air infiltration for heating $Q_{inf\text{-}H}$ and cooling $Q_{inf\text{-}C}$;
C_p is the air's specific heat capacity, which is 0.34 Wh/m³K;
G_t are annual degrees-days (kKh/yr); and
V_{inf} is the air leak rate (m³/h).

The Persily–Kronvall estimate is simple and the model is widespread in the scientific community [59]. It assumes a linear relationship between permeability at 50 Pa and average annual infiltration (Equation (4)).

$$q_{inf} = q_{50}/20, \tag{4}$$

where q_{inf} is the air permeability (m³/(h m²)).

Subsequently, this linear relationship between leakage and infiltration evolved [59] and incorporated coefficients according to the characteristics of the location of the Equations (5) and (6).

$$q_{inf} = q_{50}/N, \tag{5}$$

$$N = C \cdot cf_1 \cdot cf_2 \cdot cf_3, \tag{6}$$

where:

N is a constant;
C is the climatic factor;
cf_1 is the building's height correction factor, 1 (cf_1 = 1) to 3 (cf_1 = 0.7) floors;
cf_2 is the site screening correction factor, for well shielded cases (cf_2 = 1,2), (cf_2 = 1) (cf_2 = 1) or exposed dwellings (cf_2 = 0.9); and
cf_3 is the sealing correction factor, which depends on the value of the exponent of leak n.

This simplified extended model has been adopted to calculate the average infiltration flow in Spain's Mediterranean area, obtaining the value of climatic factor C via equivalence with US climates, based on average temperature and wind speed. For the coefficients cf_1, cf_2, and cf_3, a value equal to 1 was adopted in all three cases. The type of infiltration opening was obtained from the mean value of the flow exponent n = 0.59. The V_{inf} air leak rate, necessary to calculate the energy impact, was calculated based on the air permeability rate and the envelope surface area, as follows (Equation (7)):

$$V_{inf} = q_{inf} \cdot A_E, \tag{7}$$

where A_E is the envelope area.

Once the value of the volume of air infiltration through the envelope or air leak rate V_{inf} was obtained to calculate the energetic impact, according to the annual average corrected by means of equations (3–7), sensible and latent heats were quantified by means of the following Equations (8) and (9):

$$Q_s = V_{inf} \cdot C_e \cdot \rho \cdot (T_e - T_i), \tag{8}$$

where:

Q_s is the energy impact value of the sensible heat of the air leak rate V_{inf} (W);
V_{inf} is the air leak rate (m³/h);

C_e is the specific heat of the air under normal conditions 0.349 Wh/kgK;

ρ is the density of air (kg/m^3);

T_e is the outside air temperature in degrees Kelvin (K), considered as the average annual value; and

T_i is the air temperature inside the house (21 °C in winter and 24 °C in summer).

$$Q_l = V_{inf} \cdot C_v \cdot \rho \cdot (W_e - W_i), \tag{9}$$

where:

Q_l is the energetic impact value of the latent heat of the air leak rate V_{inf} (W);

C_v is the heat of water vaporization (0.628 W/g$_{vapour}$);

W_e is the specific humidity of the external air taken as the annual average value (g$_{vapour}$/kg);

W_i is the specific humidity of the indoor air (g$_{vapour}$/kg).

4. Results

The measurements of some thermal behaviour parameters of the materials used in the house's enclosures are presented in Table 2. They do not differ substantially from those obtained in other similar publications [40]. In this way, the thermal transmittance U values of a typical enclosure, according to Fourier or in steady state, were obtained following the values of surface thermal resistance and air chambers established by the CTE.

Table 2. Thermal conductivity values according to UNE EN ISO 10456:2012 and measurements using Mathis TCi equipment.

Materials of the Enclosure Layers	Thickness e (cm)	Thermal Conductivity λ W/m K	Thermal Conductivity λ Mathis TCi
Multilayer coating Coteterm + Calcifin	0.5	1.21	0.27
Hollow ceramic brick	11.5	0.49	0.52
SuperPan Tech P5 water-repellent wood panel	1.5	0.76	0.75
Thermal isolation. Mineral wool	14.8	0.038	0.0334
Thermal isolation. Purone mineral wool 35 QN	19.8	0.038	0.0361
Riwega USB Tape 1 PE adhesive tape	0.05	0.17	0.183
Class 3 microcement pavement	2.0	1.38	1.392
Pieces of 90 x 60 cm porcelain stoneware	1.0	2.30	2.21
Polished concrete	3.0	1.63	1.652

With these values, we proceeded to make a theoretical calculation according to Fourier and λ values extracted from the UNE EN ISO 10456: 2012 standard, of the U thermal transmittances of conventional enclosures. Table 3 shows the calculation for the opaque enclosure.

Table 3. U value of thermal transmittance for the opaque enclosure according to λ values extracted from the standard UNE EN ISO 10456:2012, established in the CTE.

	Vertical Enclosure and Horizontal Flow	Thermal Resistance		
	LAYERS	Thickness [m]	λ [W/m·K]	R [m^2·K/W]
1	Exterior environment (Rse)			0.040
2	Multilayer coating Coteterm + Calcifin	0.005	1.21	0.004
3	SuperPan Tech P5 water-repellent wood panel	0.015	0.14	0.107
4	Thermal isolation. Mineral wool type IV	0.148	0.0334	4.428
5	SuperPan Tech P5 water-repellent wood panel	0.015	0.14	0.107
6	Interior environment (Rsi)			0.130
	R_T = Sum R$_i$ [m^2·K/W]			4.816
	U_T = 1/R$_T$ [W/(m^2·K)]			0.2075

The U thermal transmittance measurements made using the sensors and Testo equipment (Figure 6) are shown in Table 4, together with the comparison of the theoretical calculation results in a steady state [36]. As shown, deviations ranging between 1% and 7% were detected. They are also compared with the usual measurement results of enclosures in standard building types in the Spanish Mediterranean area over the periods 1979–2006 and 2006–2019 [60].

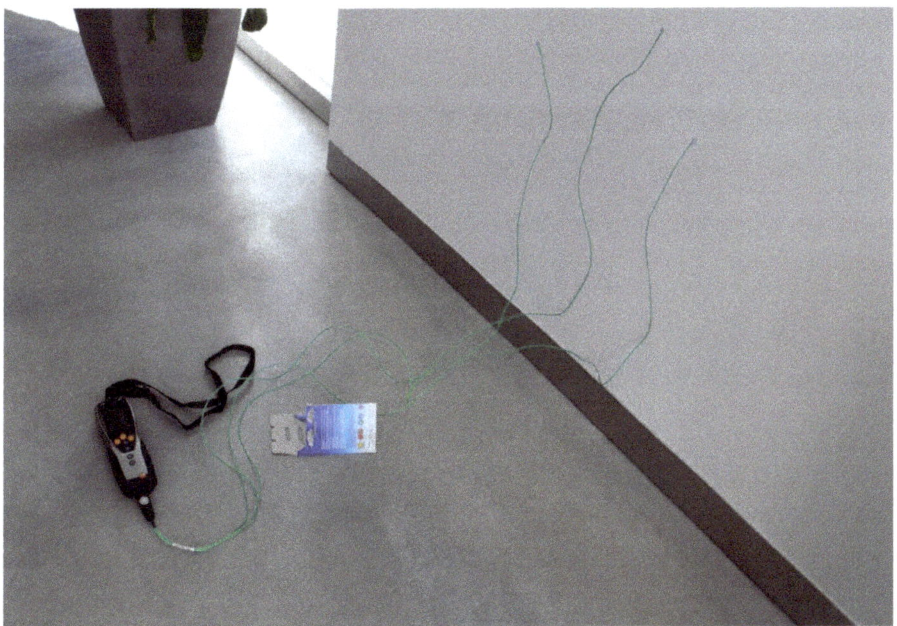

Figure 6. Measurement of U thermal transmittance of the opaque house enclosure; 11 April 2019, 5 pm.

Table 4. Value of the U thermal transmittance according to λ values extracted from the standard UNE EN ISO 10456:2012 and according to measurements made using Testo equipment. Comparison with other usual constructions.

	Passivhaus House		Standard House	
	U Value W/m^{2}K	U Value Testo Equipment W/m^{2}K	1979–2006 Period U Value W/m^{2}K	2006–2019 Period U Value W/m^{2}K
Opaque	0.207	0.221	0.476	0.421
Glazing	2.910	2.932	3.520	3.450
Roof	0.208	0.210	0.508	0.458
Ground	0.325	0.307	0.485	0.465

Regarding the air infiltration values from the Blower Door test, carried out in two phases dated 21 June 2018 and 3 October 2018 (Figure 7), the results are shown in Table 5. As can be seen, the average value is 0.51 air changes per hour (ACH) at 50 Pa, a lower value than the maximum allowed by the PH standard (Figure 8). In comparison with the values obtained for other buildings in the area, and in other geographical parts of Spain, the value is significantly lower. It is between 600% and 1200% below that of isolated single-family homes in the area built between 1979 and 2019, whose n_{50} values are between 3 and 6 ACH [61].

Table 5. Values of air infiltration through the envelope at 50 Pa. PH house compared to standard housing types in the Murcia-Alicante area.

	Passivhaus House		Standard House	
Valor n_{50} (ACH)	**Blower Door Test 1** **21 June 2018**	**Blower Door Test 2** **3 October 2018**	**1979–2006** **Period**	**2006–2019** **Period**
Depressurization	0.35	0.51	6.52	3.69
Pressurization	0.33	0.48	6.05	3.43
Average value (ACH)	0.34	0.49	6.23	3.56

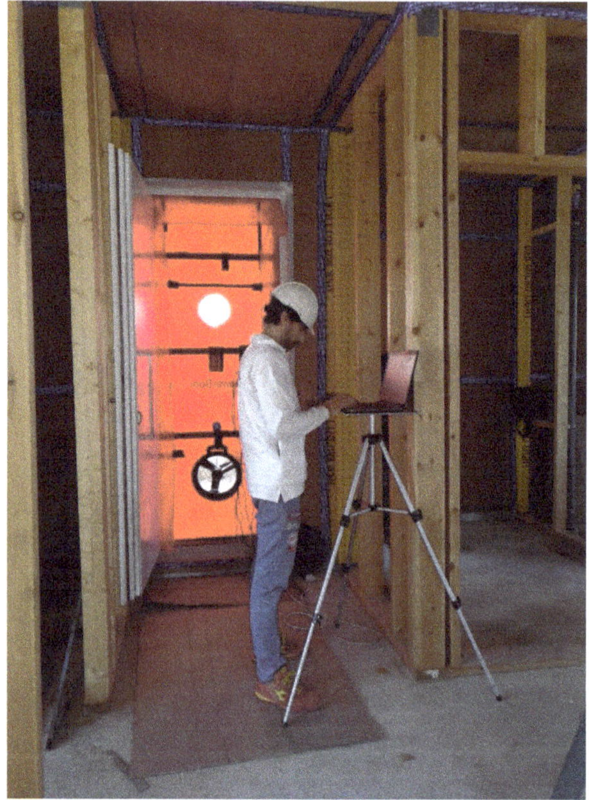

Figure 7. Image during the Blower Door test; 3 October 2018.

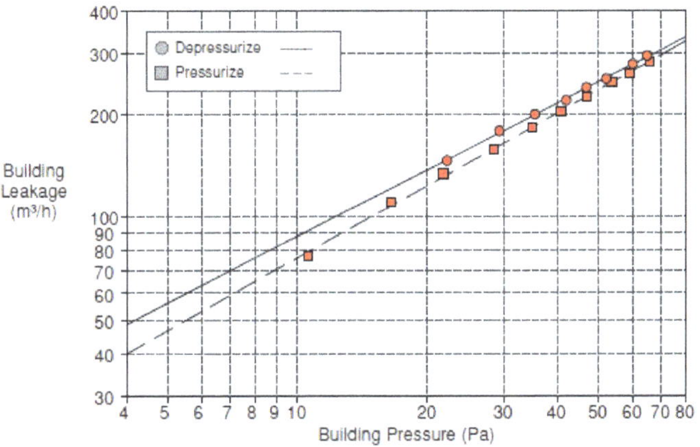

Figure 8. Results of the Blower Door test; 3 October 2018.

4.1. Energetic Impact of Heat Flow Through the Envelope

Once we obtained the enclosures' U values of thermal transmittance, the values of indoor air renewal rate according to CTE, air infiltration through the envelope, surface temperatures of the walls, house occupancy, etc., we proceeded to simulate the house's thermal behaviour using the Design Builder tool (Figure 9) [29]. It was thus possible to analyse users' feeling of comfort, the temperature gradients in the rooms, and to compare the summer and winter energy demands depending on the characteristics of the enclosures and the Blower Door test. Simulations could also be made by modifying the technical characteristics of the enclosures [62], modifying the air infiltration rate factor n_{50}, and other parameters [42]. As indicated, the interior air setpoint temperatures applied were 21 °C in winter and 24 °C in summer.

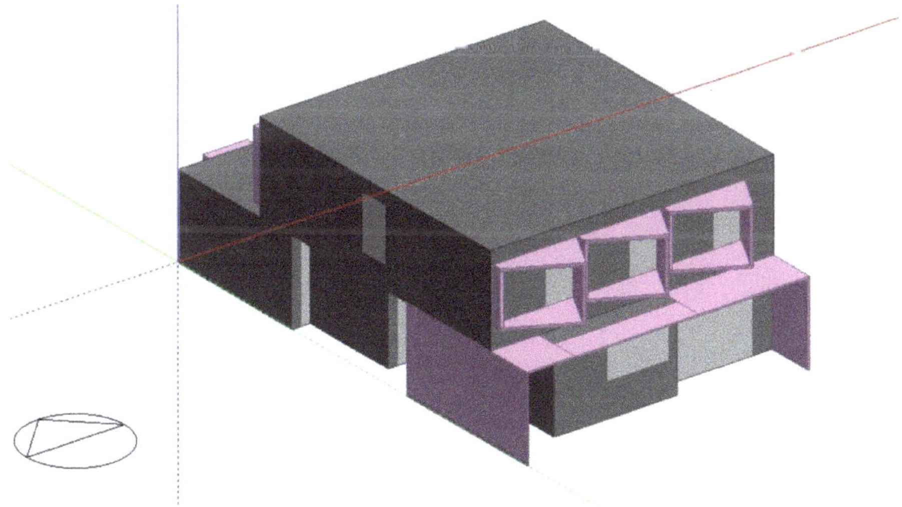

Figure 9. Model of the house in Design Builder.

The schematic diagram of the air conditioning system and the aerothermal system, as well as solar thermal panels to produce ACS, were also introduced into Design Builder. To adjust the simulation parameters to the real data, or to calibrate the model, they were compared with indoor air temperature and surface temperature values that had been obtained through monitoring. The monitoring system was designed using a wireless system. Temperature sensors were connected to small analysers ("EL-WiFi-TC. Thermocouple Probe Data logger"). The "EL-WiFi-TH, temperature and humidity data logger" analysers, with built-in sensors of temperature and relative humidity, interpreted the recorded data and sent Wi-Fi signals to a "RouterOS" router, connected to a laptop [40]. By installing the software "EasyLog Wi-Fi Software", the data was received and stored in the computer. The climate file, including external air temperatures, relative humidity, and solar radiation levels by means of a pyranometer throughout the complete one-year cycle was obtained for the house in Molina de Segura and was also introduced.

Figures 10–13 show Computational Fluid Dynamics (CFD) results obtained through the Design Builder tool for 1 August 2018 at 3 p.m. and 1 February 2019 at 9 am. The climate differences between summer and winter are remarkable. To perform the simulations, the summer period was established from 1 May to 30 November and the winter regime from 1 December to 30 April. This division was based on prior simulation results, with the resulting model calibrated according to data obtained by monitoring the complete one-year cycle [63].

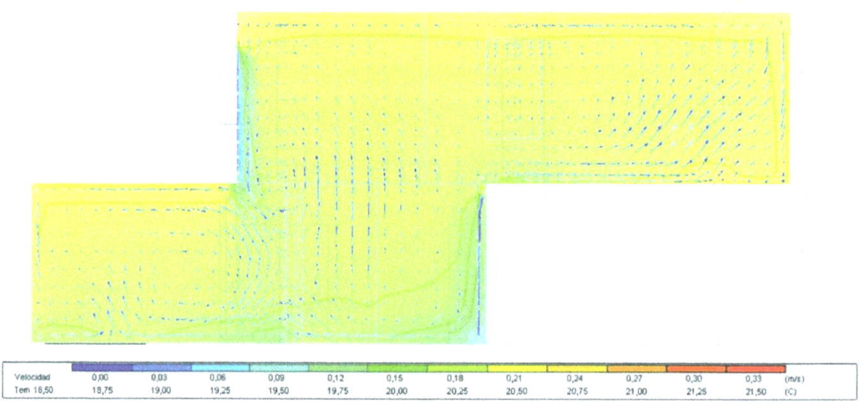

Figure 10. Simulated (CFD) of the house's thermal behaviour; 1 February 2019, 9 am.

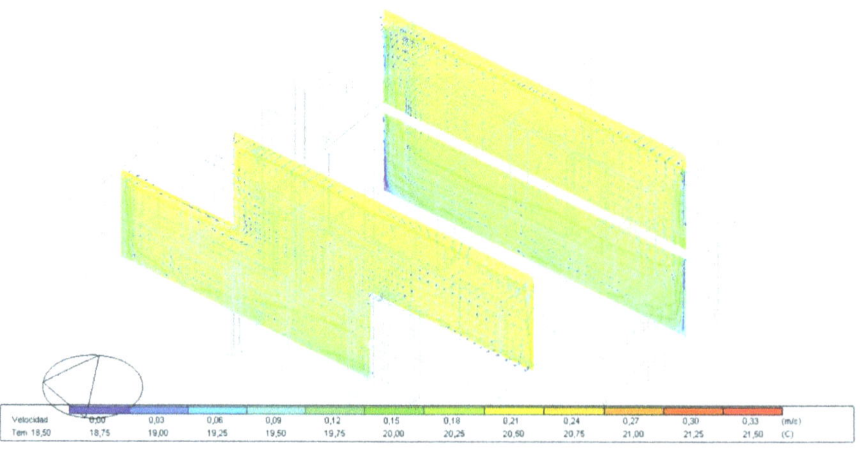

Figure 11. Simulated (CFD) of the house's thermal behaviour; 1 February 2019, 9 am.

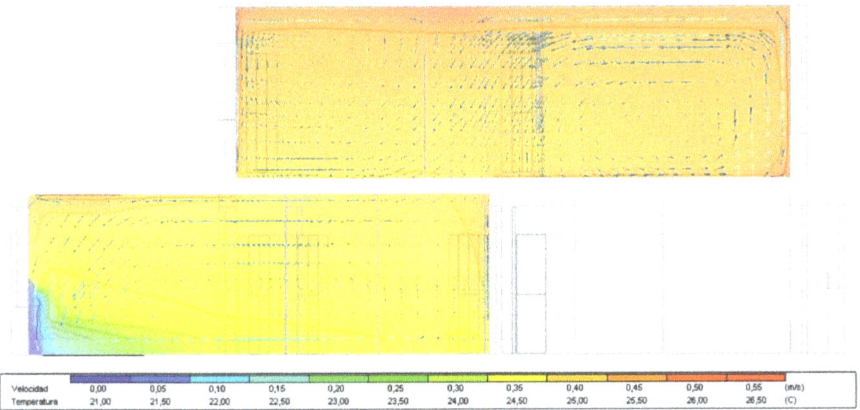

Figure 12. Simulated (CFD) of the house's thermal behaviour; 1 August 2018, 3 pm.

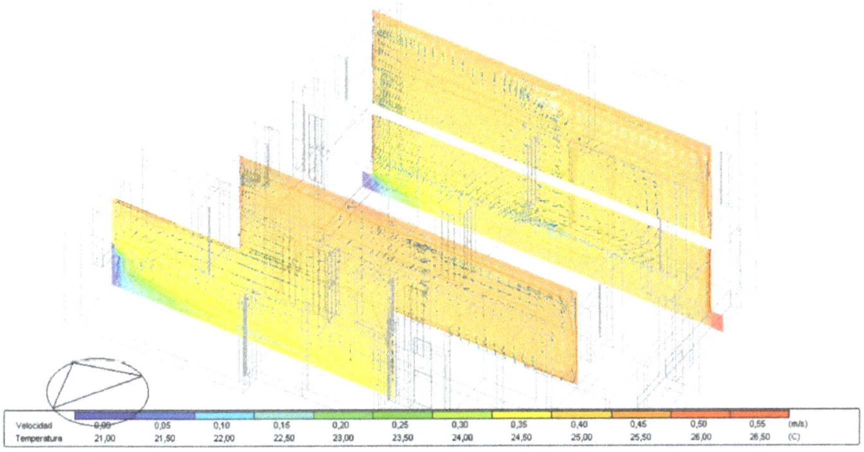

Figure 13. Simulated (CFD) of the house's thermal behaviour; 1 August 2018, 3 pm.

Based on the Design Builder simulations, impact results on annual energy demand were obtained for each housing type enclosure (Table 6). The climate file obtained in situ, by monitoring, was applied to obtain more accurate simulation results. Heat flows through the window openings accounted for almost 60% of the energy demand due to enclosure transmittance and 15% of the annual energy demand.

If we compare these results with those obtained for conventional housing constructions in the area (Table 7), the annual energy demand reduction is reduced by 39%, a high value that is mainly due to the breakage of thermal bridges and the increase in thickness of thermal insulation; 14.8 cm instead of the usual 4–6 cm [64]. The usual impact values of thermal bridges in conventional houses account for 3%–5% of the energy loss through the enclosure, while in the case of the house under study, this figure was reduced to 0.5% [65]. Simply by improving the enclosures' thermal insulation and thermal bridges, the application of the PH standard to the constructions on the Spanish Murcia-Alicante Mediterranean coast would bring about annual energy demand reductions of around 5 to 8 kWh/m^{2}year. This represents between 10% to 14% annual energy savings and this latter percentage would increase in the case of buildings built before the enforcement of the CTE.

Table 6. Energetic impact values of the different types of enclosures.

	Surface m^2	U Value W/m^{2}K	Annual Energy kWh/yr	Annual Energy Demand Impact kWh/m^{2}yr	Percentage %
Façade enclosure	181.79	0.2075	363.72	2.11	18.55
Window gap	40.09	2.91	1,124.07	6.49	57.31
Roof	173.20	0.2081	348.13	2.01	17.75
Contact with the ground	158.30	0.325	125.39	0.724	6.39
TOTAL	553.38		1,961.32	11.324	100

Table 7. Energetic impact values of the different types of enclosures.

	Passivhaus House		Standard House 1979–2006		Standard House 2006–2019	
	U Value W/m^{2}K	Annual Energy Demand Impact kWh/m^{2}yr	U Value W/m^{2}K	Annual Energy Demand Impact kWh/m^{2}yr	U Value W/m^{2}K	Annual Energy Demand Impact kWh/m^{2}yr
Opaque	0.221	2.11	0.476	4.67	0.421	4.32
Glazing	2.932	6.49	3.520	7.85	3.450	7.21
Roof	0.210	2.01	0.508	4.93	0.458	4.19
Ground	0.307	0.724	0.485	1.27	0.465	0.98
TOTAL		11.324		18.72		16.71

4.2. Energetic Impact of Air Infiltration

The energetic impact of air infiltration was obtained by applying the methodology described in Section 3.3. The values obtained are shown in Table 8. These values were also obtained in the Design Builder tool, once the value obtained in the Blower Door test was entered in the model and calibrated using the Persily–Kronvall estimate [59]. The table shows the values of the impact on annual energy demand of the air infiltration through the envelope, which alters the interior air temperature and its specific humidity [66]. The results for other conventional houses built in the same area have also been introduced. The energy impact percentages obtained are significant.

Table 8. Energetic impact values of the different types of enclosures.

	Passivhaus House	Standard House 1979–2006	Percentage	Standard House 2006–2019	Percentage
Test BD n_{50}	0.49	6.23		3.56	
Q$_s$ (kWh/m^{2}yr)	1.377	9.543		4.971	
Q$_l$ (kWh/m^{2}yr)	4.016	27.837		14.553	
Q$_t$ (kWh/m^{2}yr)	5.393	37.379	14.43%	19.524	27.62%

The energy losses due to the air infiltration in the PH house analysed were between 5 and 7 times lower, leading to significant reductions in the percentage of impact on annual energy demand. The annual energy demand was reduced between 14 and 32 kWh/m^{2}yr, in percentage terms that is between 72.38% and 85.57%. This represents a major percentage drop compared to the conventional houses on the Mediterranean Murcia-Alicante coast.

4.3. Impact of Improvements of Thermal Transmittances and Air Infiltration on Annual Energy Demand

Once the results of thermal transmission (4.1) and air infiltration (4.2) were analysed, we could lay them out in a table together with the energy demand values for summer, winter, and a year, obtained from the Design Builder tool, for each of the three types studied (Table 9).

Table 9. Comparison of the energy impact in relation to the annual energy demands of the three housing types analysed.

	PH House	Percentage	House Type 1979–2006	Percentage	House Type 2006–2019	Percentage
Energy demand in summer (kWh/m^{2}yr)	14.321		50.64		42.29	
Energy demand in winter (kWh/m^{2}yr)	12.126		35.19		29.88	
Annual energy demand (kWh/m^{2}yr)	26.447	30.81%	85.83	100%	66.57	77.56%
Energy losses through U transmission (kWh/m^{2}yr)	11.324	42.82%	18.72	21.81%	16.71	25.10%
Energy losses through infiltration Q_t (kWh/m^{2}yr)	5.393	20.39%	37.379	43.05%	19.524	29.33%

Most notably, the annual energy demand of the PH house accounted for 30.81% of the usual value for buildings in the Mediterranean region under study. Regarding the most recent buildings built under the CTE umbrella, this percentage is 39.73% [57]. A drastic drop in energy consumption can be observed. The energy losses due to enclosure transmittance and air infiltration in the PH house were much lower, i.e., between 21.5 and 46 kWh m^{2}yr. They represent major annual energy demand drops for conventional housing on the Mediterranean Murcia-Alicante coast. The energy losses by infiltration were also found to be around 30%–40% of conventional housing energy demands, while this value was reduced to 20% in the case of the PH.

If we consider the values obtained in other studies, according to which the average impact of air infiltration on annual energy demand is around 12 kWh/m^{2}yr for Alicante, 10 kWh/m^{2}yr for Malaga, or 17 kWh/m^{2}yr for Barcelona [61], we can conclude that by applying the PH standard and its technical construction requirements, annual energy demand due to infiltration could be reduced by between 100% and 300%.

5. Conclusions

Quantifying the energetic impact of thermal transmission through enclosures and air infiltration is decisive to adapt the Passivhaus standard to the Spanish Mediterranean region. When gaps and glazing are designed to protect against solar radiation, overheating of indoor air and interior surfaces can be avoided, making it easier to implement the standard. This design layout leads to significant reductions in annual energy demand. Very thick thermal insulation could thus be applied to the envelope, which would work very well in winter, with energy demands below 15 kWh/m^{2}year and optimal comfort conditions with a well-designed air conditioning system in summer. The annual energy demand would not be too high, below 30 kW/m^{2}yr, a much lower value than the usual standards. Heat recovery systems relating to air renewal in summer, combined with the VRV system's air dehumidification, allow us to conclude the implementation of the standard could be generalised in the area, with low energy consumption. Low air infiltration implies low dehumidification system consumption, maintaining 50% of RH.

Highly satisfactory energy savings were obtained for the PH house under study, as follows:

- Energy losses through enclosure transmittance were 11.324 kWh/m^{2}yr, between 32% and 40% lower than those obtained in the usual homes in the region;
- Losses due to air infiltration were very low, of 5.393 kWh/m^{2}yr, given that the Blower Door test produced a very low n_{50} value of 0.49 ACH. They were between 5 and 7 times lower than in conventional homes in the area;
- The annual energy demand of PH housing was 69.19% below the usual value for buildings in the Mediterranean region. Regarding the most recent buildings built under the CTE umbrella, the reduction was 60.27%.

In future studies, investments necessary to build the house will be quantified and compared to conventional Murcia-Alicante coast standards. The energy saving values obtained in this study will be applied and the investments' amortisation period will be calculated. It will be possible to demonstrate the adequacy of the Passivhaus standard to the Administrations and the necessary corrections to apply the standard under the Spanish Mediterranean region climate. Policies proposing incentives to implement these construction systems and specific installations would undoubtedly lead to significant energy savings and thus reduce the environmental impacts deriving from the building and use of residential developments.

Author Contributions: Conceptualisation, methodology, measurements of *U*-value and test Blower Door, V.E.-I.; investigation, writing—original draft preparation, V.E.-I.; control of the data collecting of the monitorisation system and graphics, C.S.-S.; simulations by Design Builder, A.E.-F.; calibration of the model in Design Builder, and interpretation of data results, V.E.-I.; energetic impact of air infiltration, R.P.-G. and V.E.-I.

Funding: This research has been funded by the Vicerectorate of Campus Facilities & Techonology of the Universidad de Alicante, ref. 2018 34406P0002 P102 68000 (VIGROB 223).

Acknowledgments: Our thanks to the company "Habitar Natural 100 por 100 madera S.L." for their help in this study and to Jorge Vera Morales, promoter of the work, for having allowed us to access the house, monitor it, and take measurements. Our gratitude to Joaquín Ruiz Piñera, architect, designer, and director of the work, and Asier Elorza Echebarría, director of execution of the work, for their help in technical aspects.

Conflicts of Interest: The authors declare no conflict of interest.

References

1. European Commission. Directive 2002/91/EC of the European Parliament andof the Council of 16 December 2002 on the energy performance of buildings. *Off. J. Eur. Commun.* **2003**, *L1*, 65–70.

2. European Commission. Directive 2010/31/EU of European Parliament and ofthe Council of 19 May 2010 on the energy performance of buildings (recast). *Off. J. Eur. Union* **2010**, *L153*, 13–35.

3. Attia, S.; Eleftheriou, P.; Xeni, F.; Morlot, R.; Ménézo, C.; Kostopoulos, V.; Betsi, M.; Kalaitzoglou, I.; Pagliano, L.; Cellura, M.; et al. Overview and future challenges of nearly zero energy buildings (nZEB) design in Southern Europe. *Energy Build.* **2017**, *155*, 439–458. [CrossRef]

4. European Commission. Directive 2012/27/EU of European Parliament and of the Council of 25 October 2012 on energy efficiency, amending Directives2009/125/EC and 2010/30/EU and repealing Directives 2004/8/EC and2006/32/EC. *Off. J. Eur. Union* **2012**, *L315*, 1–56.

5. Technical Building Code CTE. *Basic Document on Energy Savings DB-HE*; Spanish Ministry of Housing: Madrid, Spain, 2006. (In Spanish)

6. *Orden VIV/984/2009, de 15 de abril, por la que se modifican determinados documentos básicos del Código Técnico de la Edificación aprobados por el Real Decreto 314/2006, de 17 de marzo, y el Real Decreto 1371/2007, de 19 de octubre;* Spanish Ministry of Housing: Madrid, Spain, 2009. (In Spanish)

7. *Orden FOM/1635/2013, de 10 de septiembre, por la que se actualiza el Documento Básico DB-HE Ahorro de Energía, del Código Técnico de la Edificación, aprobado por Real Decreto 314/2006, de 17 de marzo; (Publicado en: «BOE» núm. 219, de 12 de septiembre de 2013, páginas 67137–67209);* Spanish Ministry of Public Works and Transport: Madrid, Spain, 2013. (In Spanish)

8. Passive House —Passivhaus Institut (PHI). Available online: http://passivehousecom/02informations/01whatisapassivehouse/01whatisapassivehousehtm (accessed on 9 April 2019).

9. Consoli, A.; Costanzo, V.; Evola, G.; Marletta, L. Refurbishing an existing apartment block in Mediterranean climate: Towards the Passivhaus standard. *Energy Procedia* **2017**, *111*, 397–406. [CrossRef]

10. Taleb, H.M. Using passive cooling strategies to improve thermal performance and reduce energy consumption of residential buildings in U.A.E. buildings. *Front. Arch. Res.* **2014**, *3*, 154–165. [CrossRef]

11. Schnieders, J.A.; Hermelink, A. CEPHEUS results: Measurements and occupants' satisfaction provide evidence for Passive Houses being an option for sustainable building. *Energy Policy* **2006**, *34*, 151–171. [CrossRef]

12. Costanzo, V.; Fabbri, K.; Piraccini, S. Stressing the passive behavior of a Passivhaus: An evidence-based scenario analysis for a Mediterranean case study. *Build. Environ.* **2018**, *142*, 265–277. [CrossRef]

13. Intelligent Energy Europe, Passive-On Project: Towards Passive Homes. 2007. Available online: http://dx.doi.org/10.1017/CBO9781107415324.004 (accessed on 11 April 2019).

14. Koller, C.; Talmon-Gros, M.J.; Junge, R.; Schuetze, T. Energy Toolbox—Framework for the Development of a Tool for the Primary Design of Zero Emission Buildings in European and Asian Cities. *Sustainability* **2017**, *9*, 2244. [CrossRef]

15. MED-ENEC. *Energy Efficiency in the Construction Sector in the Mediterranean*; MED-ENEC: Cairo, Egipt, 2011; Available online: http://www.med-enec.com (accessed on 18 April 2019).

16. IEEA. The Passivehaus Standard in European Warm Climates, EC Funded Project: Marketable Passive Homes for Winter and Summer Comfort. 2007. Available online: http://www.eerg.it/passive-on.org/CD/1.%20Technical%20Guidelines/Part%203/Part%203.pdf (accessed on 17 April 2019).

17. Algerian Ministry of Energy and Mines. Renewable Energy and Energy Efficiency Algerian Program. *Renew. Sustain. Energy Rev.* **2011**, *13*, 1584–1591.

18. Schnieders, J. Passive Houses in South West Europe: A Quantitative Investigation of Some Passive and Active Space Conditioning Techniques for Highly Energy Efficient Dwellings in the South West European. Region. Dissertation, Technical University Kaiserslautern, Kaiserslautern, Germany, 2009.

19. Figueiredo, A.; Figueira, J.; Vicente, R.; Maio, R. Thermal comfort and energy performance: Sensitivity analysis to apply the Passive House concept to the Portuguese climate. *Build. Environ.* **2016**, *103*, 276–288. [CrossRef]

20. Schnieders, J.; Wolfgang Feist, W.; Rongen, L. Passive Houses for different climate zones. *Energy Build.* **2015**, *105*, 71–87. [CrossRef]

21. EN 15251. *Indoor Environmental Input Parameters for Design and Assessment of Energy Performance of Buildings Addressing Indoor Air Quality, Thermal Environment, Lighting and Acoustic*; European Committee for Standardization: Brussels, Belgium, 2008.

22. Suárez, I.; Prieto, M.M.; Salgado, I. Dynamic evaluation of the thermal inertia of a single-family house: Scope of the retrofitting requirements to comply with Spanish regulations. *Energy Build.* **2017**, *153*, 209–218.

23. Pesic, N.; Roset Calzada, J.; Muros Alcojor, A. Natural ventilation potential of the Mediterranean coastal region of Catalonia. *Energy Build.* **2018**, *169*, 236–244. [CrossRef]

24. Fokaides, P.A.; Christoforou, E.; Ilic, M.; Papadopoulos, A. Performance of a Passive House under subtropical climatic conditions. *Energy Build.* **2016**, *133*, 14–31. [CrossRef]

25. Ali-Toudert, F.; Weidhaus, J. Numerical assessment and optimization of a low-energy residential building for Mediterranean and Saharan climates using a pilot project in Algeria. *Renew. Energy* **2017**, *101*, 327–346. [CrossRef]

26. Altan, H.; Gasperini, N.; Moshaver, S.; Frattari, A. Redesigning Terraced Social Housing in the UK for Flexibility Using Building Energy Simulation with Consideration of Passive Design. *Sustainability* **2015**, *7*, 5488–5507. [CrossRef]

27. Mihai, M.; Tanasiev, V.; Dinca, C.; Badea, A.; Vidu, R. Passive house analysis in terms of energy performance. *Energy Build.* **2017**, *144*, 74–86. [CrossRef]

28. Harkouss, F.; Fardoun, F.; Biwole, P.H. Passive design optimization of low energy buildings in different climates. *Energy* **2018**, *165*, 591–613. [CrossRef]

29. Dan, D.; Tanasa, C.; Stoian, V.; Brata, S.; Stoian, D.; Nagy Gyorgy, T.; Florut, S.C. Passive house design—An efficient solution for residential buildings in Romania. *Energy Sustain. Dev.* **2016**, *32*, 99–109. [CrossRef]

30. Design Builder. *Design Builder EnergyPlus Simulation Documentation for Design Builder v. 4.5.*; Design Builder Sortware Ltd.: Stroud, UK, 2015.

31. Mlakar, J.; Strancar, J. Overheating in residential passive house: solution strategies revealed and confirmed through data analysis and simulations. *Energy Build.* **2011**, *43*, 1443–1451. [CrossRef]

32. Presidencia de Gobierno. *Real Decreto 2429/1979, de 6 de julio, por el que se Aprueba la Norma Básica de Edificación NBE-CT-79, sobre Condiciones Térmicas en los Edificios*; Boletín Oficial del Estado: Madrid, Spain, 1979; Volume 253, pp. 24524–24550.

33. Bruno, R.; Bevilacqua, P.; Cuconati, T.; Arcuri, N. Energy evaluations of an innovative multi-storey wooden near Zero Energy Building designed for Mediterranean areas. *Appl. Energy* **2019**, *238*, 929–941. [CrossRef]

34. Kottek, M.; Grieser, J.; Beck, C.; Rudolf, B.; Rubel, F. World Map of the Köppen-Geiger climate classification updated. *Meteorol. Z.* **2006**, *15*, 259–263. [CrossRef]

35. ure(2017). World Map of the Köppen-Geiger Climate Classification Updated. High Resolution Map and Data (Version March 2017). KMZ File for Google Earth (High Res): Global_1986-2010_KG_5m.kmz.

Climate Change & Infectious Diseases. Available online: http://koeppen-geiger.vu-wien.ac.at (accessed on 13 April 2019).

36. Echarri, V.; Espinosa, A.; Rizo, C. Thermal Transmission through Existing Building Enclosures: Destructive Monitoring in Intermediate Layers versus Non-Destructive Monitoring with Sensors on Surfaces. *Sensors* **2017**, *17*, 2848. [CrossRef] [PubMed]

37. Dong, M.N.; Lu, Z.; Mo, T.Z.; Yang, J.Y.; Leng, Y.F.; Yang, L.L. Quantitative analysis on the effect of thermal bridges on energy consumption of residential buildings in hot summer and cold winter region. *J. Chongqing Jianzhu Univ.* **2008**, *30*, 5–8.

38. Garay, R.; Uriarte, A.; Apraiz, I. Performance assessment of thermal bridge elements into a full scale experimental study of a building façade. *Energy Build.* **2014**, *85*, 579–591. [CrossRef]

39. Wang, S.; Chen, Y. A simple procedure for calculating thermal response factors and conduction transfer functions of multilayer walls. *Appl. Therm. Eng.* **2002**, *22*, 333–338. [CrossRef]

40. Echarri Iribarren, V.; Galiano Garrigós, A.L.; González Avilés, A.B. Ceramics and healthy heating and cooling systems: Thermal ceramic panels in buildings. Conditions of comfort and energy demand versus convective systems. *Informes de la Construcción* **2016**, *68*, 19–32.

41. ISO. 13786:2007. Thermal Performance of Building Components. Dynamic Thermal Characteristics. Calculation Methods. Available online: https://www.iso.org/standard/40892.html (accessed on 19 September 2018).

42. Aznar, F.; Echarri, V.; Rizo, C.; Rizo, R. Modelling the Thermal Behaviour of a Building Facade Using Deep Learning. *PLoS ONE* **2018**, *13*, e0207616. [CrossRef]

43. CTE. Código Técnico de la Edificación. R/D 314/2006, de 17 de Marzo. Available online: http://www.codigotecnico.org/images/stories/pdf/realDecreto/RD3142006.pdf (accessed on 9 September 2017).

44. UNE EN ISO. 10456:2012. Materiales y Productos Para la Edificación. Propiedades Higrotérmicas. Valores Tabulados de Diseño y Procedimientos Para la Determinación de los Valores Térmicos Declarados y de Diseño. 2012. Available online: http://www.aenor.es/aenor/normas/normas/fichanorma.asp?tipo=N&codigo=N0049362#.Wgbq4tThCmw (accessed on 9 October 2017).

45. International Organization for Standardization. *Building Components and Building Elements. Thermal Resistance and Thermal Transmittance. Calculation Method*; ISO Standard 6946; International Organization for Standardization: Geneva, Switzerland, 2007; Available online: https://www.iso.org/obp/ui/#iso:std:iso:6946:ed-2:v1:en (accessed on 13 April 2019).

46. Evangelisti, L.; Battista, G.; Guattari, C.; Basilicata, C.; de Lieto Vollaro, R. Influence of the thermal inertia in the European simplified procedures for the assessment of buildings' energy performance. *Sustainability* **2014**, *6*, 4514–4524. [CrossRef]

47. HULC. Herramienta unificada Lider-Calener. Orden FOM/1635/2013, de 10 de septiembre (BOE de 12 de septiembre), por la que se actualiza el Documento Básico DB HE «Ahorro de Energía», del CTE. Available online: https://www.codigotecnico.org/index.php/menu-recursos/menu-aplicaciones/282-herramientaunificada-lider-calener.html (accessed on 9 September 2017).

48. Andújar Márquez, J.M.; Martínez Bohórquez, M.A.; Gómez Melgar, S. A New Metre for Cheap, Quick, Reliable and Simple Thermal Transmittance (U-Value) Measurements in Buildings. *Sensors* **2017**, *17*. [CrossRef] [PubMed]

49. Deconinck, A.H.; Roels, S. Comparison of characterisation methods determining the thermal resistance of building components from onsite measurements. *Energy Build.* **2016**, *130*, 309–320. [CrossRef]

50. ISO. 9869-1:2014. *Thermal Insulation-Building Elements-In Situ Measurement of Thermal Resistance and Thermal Transmittance. Part 1.: Heat Flow Meter Method*; ISO: Geneva, Switzerland, 2014; Available online: https://www.iso.org/standard/59697.html (accessed on 13 April 2019).

51. Gaspar, K.; Casals, M.; Gangolells, M. A comparison of standardized calculation methods for in situ measurements of façades U-value. *Energy Build.* **2016**, *130*, 592–599. [CrossRef]

52. Albatici, R.; Tonelli, A.M.; Chiogna, M. A comprehensive experimental approach for the validation of quantitative infrared thermography in the evaluation of building thermal transmittance. *Appl. Energy* **2015**, *141*, 218–228. [CrossRef]

53. Tinti, A.; Tarzia, A.; Passaro, A.; Angiuli, R. Thermographic analysis of polyurethane foams integrated with phase change materials designed for dynamic thermal insulation in refrigerated transport. *Appl. Therm. Eng.* **2014**, *70*, 201–210. [CrossRef]

54. Theodosiou, T.G.; Tsikaloudaki, A.G.; Kontoleon, K.J.; Bikas, D.K. Thermal bridging analysis on cladding systems for building facades. *Energy Build.* **2015**, *109*, 377–384. [CrossRef]

55. Feijó-Muñoz, J.; Poza-Casado, I.; González-Lezcano, R.A.; Pardal, C.; Echarri, V.; Assiego de Larriva, R.; Fernández-Agüera, J.; Dios-Viéitez, M.J.; del Campo-Díaz, V.J.; Montesdeoca Calderín, M.; et al. Methodology for the study of the envelope airtightness of residential buildings in Spain: A case study. *Energies* **2018**, *11*, 704. [CrossRef]

56. Bienvenido-Huertas, D.; Fernández Quiñones, J.A.; Moyano, J.; Rodríguez-Jiménez, C.E. Patents Analysis of Thermal Bridges in Slab Fronts and Their Effect on Energy Demand. *Energies* **2018**, *11*, 2222. [CrossRef]

57. Echarri-Iribarren, V.; Rizo-Maestre, C.; Echarri-Iribarren, F. Healthy Climate and Energy Savings: Using Termal Ceramic Panels and Solar Thermal Panels in Mediterranean Housing Blocks. *Energies* **2018**, *11*, 2707. [CrossRef]

58. Younes, C.; Shdid, C.A.; Bitsuamlak, G. Air infiltration through building envelopes: a review. *J. Build. Phys.* **2012**, *35*, 267–302. [CrossRef]

59. Sherman, M.H. Estimation of infiltration from leakage and climate indicators. *Energy Build.* **1987**, *10*, 81–86. [CrossRef]

60. Echarri, V. Thermal Ceramic Panels and Passive Systems in Mediterranean Housing: Energy Savings and Environmental Impacts. *Sustainability* **2017**, *9*, 1613. [CrossRef]

61. Feijó-Muñoz, J.; Pardal, C.; Echarri, V.; Fernández-Agüera, J.; Assiego de Larriva, R.; Montesdeoca Calderín, M.; Poza-Casado, I.; Padilla-Marcos, M.A.; Meiss, A. Energy impact of the air infiltration in residential buildings in the Mediterranean area of Spain and the Canary islands. *Energy Build.* **2019**, *188–189*, 226–238. [CrossRef]

62. Bienvenido-Huertas, D.; Moyano, J.; Rodríguez-Jiménez, C.E.; Marín, D. Applying an artificial neural network to assess thermal transmittance in walls by means of the thermometric method. *Appl. Energy* **2019**, *233–234*, 1–14. [CrossRef]

63. Echarri-Iribarren, V.; Rizo-Maestre, C.; Sanjuan-Palermo, J.L. Underfloor Heating Using Ceramic Thermal Panels and Solar Thermal Panels in Public Buildings in the Mediterranean: Energy Savings and Healthy Indoor Environment. *Appl. Sci.* **2019**, *9*, 2089. [CrossRef]

64. Šadauskiene, J.; Ramanauskas, J.; Šeduikyte, L.; Daukšys, M.; Vasylius, A. A Simplified Methodology for Evaluating the Impact of Point Thermal Bridges on the High-Energy Performance of a Passive House. *Sustainability* **2015**, *7*, 16687–16702. [CrossRef]

65. Fu, X.; Qian, X.; Wang, L. Energy Efficiency for Airtightness and Exterior Wall Insulation of Passive Houses in Hot Summer and Cold Winter Zone of China. *Sustainability* **2017**, *9*, 1097. [CrossRef]

66. Johnston, D.; Siddall, M. The Building Fabric Thermal Performance of Passivhaus Dwellings—Does It Do What It Says on the Tin? *Sustainability* **2016**, *8*, 97. [CrossRef]

Article

Passive Design Strategies for Residential Buildings in Different Spanish Climate Zones

Maria-Mar Fernandez-Antolin [1,*], José Manuel del Río [1], Vincenzo Costanzo [2], Francesco Nocera [3] and Roberto-Alonso Gonzalez-Lezcano [1]

1 Escuela Politécnica Superior, Universidad CEU San Pablo, Montepríncipe Campus, 28668 Boadilla del Monte, Madrid, Spain
2 Department of Electric, Electronic and Computer Engineering, University of Catania, Via Santa Sofia, 6495123 Catania, Italy
3 Department of Civil Engineering and Architecture, University of Catania, Via Santa Sofia, 6495123 Catania, Italy
* Correspondence: mariamar.fernandezantolin@ceu.es; Tel.: +34-(0)6-1618-7449

Received: 23 July 2019; Accepted: 30 August 2019; Published: 4 September 2019

Abstract: The Passive House (PH) concept is considered an efficient strategy to reduce energy consumption in the building sector, where most of the energy is used for heating and cooling applications. For this reason, energy efficiency measures are increasingly implemented in the residential sector, which is the main responsible for such a consumption. The need for professionals dealing with energy issues, and particularly for architects during the early stages of their architectural design, is crucial when considering energy efficient buildings. Therefore, architects involved in the design and construction stages have key roles in the process of enhancing energy efficiency in buildings. This research work explores the energy efficiency and optimized architectural design for residential buildings located in different climate zones in Spain, with an emphasis on Building Performance Simulation (BPS) as the key tool for architects and other professionals. According to a parametric analysis performed using Design Builder, the following optimal configurations are found for typical residential building projects: North-to-South orientation in all the five climate zones, a maximum shape factor of 0.48, external walls complying with the maximum U-value prescribed by Spanish Building Technical Code (0.35 $Wm^{-2}K^{-1}$) and a Window-to-Wall Ratio of no more than 20%. In terms of solar reflectance, it is found that the use of light colors is better in hotter climate zones A4, B4, and C4, whereas the best option is using darker colors in the colder climate zones D3 and E1. These measures help reaching the energy demand thresholds set by the Passivhaus Standard in all climate zones except for those located in climates C4, D3 and E1, for which further passive design measures are needed.

Keywords: energy efficiency; dynamic building simulation; passive house; building energy performance; passive strategies

---

1. Introduction

Nowadays, a wide consensus has been achieved upon the importance of good architectural design and its relationship with energy consumption [1]. As a matter of fact, energy efficiency in the residential sector is one of the priority objectives of the European Union, which in the Directive 2010/31/EU requires 20% reduction in emissions of global warming gases, 20% reduction in energy consumption, and 20% increment in the use of renewable energies with respect to 1990 levels [2].

It is estimated that energy savings up to 27% will be achieved in residential buildings by 2020 as per Directive 2012/27/EU and Directive 2018/844/UE [3,4]. The impact of buildings is significant, and amounts to approximately 40% of global energy consumption and one third of global greenhouse gas (GHG) emissions [5].

When it comes to the construction of dwellings in the Spanish Mediterranean area, what is found is a general and widespread lower energy efficiency than other countries within the European Union facing the Mediterranean basin [6].

In typical Spanish dwellings, energy consumption due to heating and cooling represents approximately 48% of the total energy used [7]. To help architects and designers reduce the operational energy consumption of buildings, the Spanish Technical Building Code (CTE) recommends the application of a tool called LIDER-CALENER GT (HULC) [8,9]. Such a tool is used to verify the compliance of the project with the prescriptions set by CTE's basic document DB-HE, as well as for evaluating the energy demand with respect to the DB-HE1's requirements (item 2.2.1) regarding the limitation of the energy demand in residential buildings.

In the light of the above, the Passive House standard is rapidly spreading all over the world with approximately 25,000 passive houses in use worldwide [10]. Several authors [11–13] claim that they can save up to 50% of the total primary energy consumption. A comparison between certified passive houses and generic low-energy houses revealed that passive house CO_2 emissions were approximately 25–40% lower, with only a 5% of increase in initial construction costs [14].

Various studies indicate that passive design measures and orientation have a considerable influence on energy efficiency, comfort and safety. There are several researches work whose objective is to discuss the implementation of passive strategies in order to reduce the energy demand in buildings [15–17].

An adequate practice of passive design involves several aspects of building design [18,19] such as the orientation of the main façades and windows, the choice of appropriate walls' materials, thermal insulation and Window-to-Wall Ratio (WWR), along with the design of shading devices and the implementation of natural ventilation techniques [20–24].

The PH concept furthers the traditional passive building approach by improving thermal comfort conditions at minimum energy costs [25]. The distinctive features of the PH standard, as prescribed in the updated official website [26,27], are:

- Space heating demand: not to exceed 15 kWhm^{-2} annually or not to exceed 10 Wm^{-2} of peak demand;
- Space cooling demand: matches the heat demand requirements with an additional, climate—dependent allowance for dehumidification;
- Primary energy demand: not to exceed 120 kWhm^{-2} annually for all domestic applications (heating, cooling, hot water and domestic electricity);
- Air tightness: air leakages should be kept below 0.6 air changes per hour at 50 Pa pressure difference (to assess on—site through a blower—door test).
- Thermal comfort for summer operation: a maximum of 10% of hours in a year exceeding 25 °C can be accepted (overheating criterion);
- High thermal insulation values in all components of the building envelope (typical U values between 0.6 and 0.15 Wm^{-2}K^{-1}).

The vast majority of passive structures have been built in northern and central European countries, but there is a significant interest for research and development targeted to the adaptation of passive houses under different climatic conditions, especially for the Mediterranean climate. In this sense, the standard allows for exceeding the overheating temperature threshold of 25 °C for no more than 10% of the cooling period in warmer climates [28]. The main problem when adapting buildings on the Mediterranean coast to the PH standard is that cooling needs must be taken into great account in summer in addition to heating demands in winter [29]. This research aims to highlight the effectiveness of the application of passive house principles on new buildings in the early design stage of residential buildings projects in the Mediterranean zone. Applying the principles of passive design to new buildings costs little or nothing while also responding to local climate and site conditions to maximize building users' comfort and health while minimizing energy use. Architects, designers, builders and stakeholders have to be aware of the opportunity raising from incorporating the principles of passive houses in the early stages of a project. Moreover, the large number of opportunities in the market today makes necessary for architects and designers to have a tool to assist them in identifying the best combinations for any specific situation [30] in order to determine which design strategy should be pursued more actively to achieve better energy performance. In this context, energy simulation is a powerful tool to improve building design [31]. The use of Building Performance Simulation Tools (BPSTs) offers the architects and designers In general a thorough understanding of the impact of several design variables [32]. This sort of simulations cannot be carried out using semi steady state tool like the Passive House Planning Package (PHPP, [33]) when it comes to consider a building's summer behavior and thus inertial effects. With this in mind, parametric energy dynamic simulations were performed to evaluate the effectiveness of passive house measures on new building design located in Spain with different climate conditions.

Despite the importance of using BPSTs, most architects and designers committed to design passive and low-energy buildings still do not use such tools, but rather rely on basic rules and environmental design guidelines [34–36]. Due to this reason, communication with software developers is often deficient [37]. Technical and non-technical barriers have been detected that prevent adopting BPSTs in the context of design, and producing a usage gap that will only begin to be solved when users will be prepared to properly interpret the results of simulations [38]. Since the late 1990s, software developers and researchers have tried to provide efficient visualization systems in which designers can easily compare and evaluate alternatives along with their proposals [39,40]. In order to understand the results of a simulation and make design decisions based on them, there is a need to research these aspects of architectural design [41,42]. There are examples of data visualizations which relate the decision-making process of design with objectives [43]. A powerful and meaningful way to explore several design solutions, and their impact on the resulting operational energy demand, is given by a parametric analysis of selected key variables [44]. The parametric sensitivity analysis is used to understand how the input parameters propagate through the model [45]; it is usually accomplished by assigning ranges of values in order to "evaluate the influence or relative importance of each input and output" [46]. In order to facilitate the interpretation of data, users must be provided with the ability to compare and relate information [47]. In the following research work, two main objectives are taken into account. The first objective is to provide building designers with suggestions for the achievement of low-energy building design within the five climate zones in Spain through the use of passive strategies. The second one is to help designers understand the complexity of the simulation outcomes through graphic representations that relates the results of the simulations with architectural design decisions.

2. Methodology

The research aims to highlight the effectiveness of the application of passive house principles on new buildings. To this purpose, parametric energy dynamic simulations were performed to evaluate the effectiveness of passive house measures on new buildings located in Spain under different climate conditions.

In this section, the different types of buildings representative of typical house dwellings in Spain are first introduced and discussed. Then, the design parameters taken into account for running a parametric analysis of different design options are presented. Finally, the different climate conditions pertaining to several climate zones where the buildings are simulated are explained.

2.1. Building Types and Simulation Parameters

In order to consider the building types that are most representative of the residential building stock in Spain, the classification resulting from the analysis of statistical data obtained from the Rehenergía project [48] and from the Spanish National Statistics Institute (INE data base [49]) has been used, with a focus on multi-dwelling units (MDUs). This is why the resulting simulation space does not take into account further variations possible through a brute-force parametric approach.

Table 1 shows the building types analyzed in this research. They are the AT1b with a prominent linear shape, and the AT2 and AT3 types that are both cubic and hollow but show different shape factors (i.e., different ratios of the heat transfer surface to the enclosed volume).

Table 1. Building types analyzed.

Building Types	Dimensions (m)	Volume (m^3)	A_{ext} (m^2)	Shape Factor (—)
Building A Linear without hollow volumes, four storey, with an average dwelling area of 80 m^2 and height of 2.8 m	AT1b (60 × 15 × 11.20 m)	10080	2580	0.26
Building B Tower with a hollow volume, four storey, with an average dwelling area of 80 m^2 and height of 2.8 m	AT2 (30 × 30 × 11.20 m)	9531.20	2508.60	0.26
Building C Cubic building with a hollow volume, three storey, with an average dwelling area of 80 m^2 and height of 2.8 m	AT3 (15 × 15 × 8.40 m)	1755.60	847.40	0.48

As mentioned in the introduction, the study is based on dynamic simulations carried out using the Energy Plus software through the Design Builder graphical interface [50]. A total number of 300 simulations are accomplished. A first set of simulations-for a total number of 240 models-considers the parametric variations of climate zones (5), building types (3), building orientation (4), U-values of external walls (2) and their external solar reflectance values (2). For such group of simulations, the parameter held constant is the Window-to-Wall Ratio (WWR), for which a constant value of 20% has been used for all facades. On the other hand, the second set of simulations includes 60 models and considers the parametric variations of climate zones (5), building types (3) and building orientation (4). The parameters considered constant in this case are the WWR (40% for all the facades), the U-value for the external walls (U = 1.58 Wm^{-2}K^{-1}) and walls solar reflectance (equal to 0.64). All these variations are summarized in Figure 1.

Table 2 summarizes all the simulation parameters, mostly referring to the Spanish regulations in force [6]. Simulations are carried out using the set back and set point temperature schedules presented in Table 3, i.e., heating and cooling ideal loads air systems are supposed to be in use whenever needed for keeping indoor air temperature within the temperature ranges set (17 to 20 °C in winter and 25 to 27 °C in summer respectively).

Table 2. Parameters variation assessed in the simulations.

Design Parameters	Abbreviations
Building type	A, B, C
Building rotation with respect to the North (°)	0°, 60°, 90°, 120°
U—values of the external walls (Wm^{-2}K^{-1})	$E_1 = 1.58$, $E_2 = 0.35$
Solar reflectance (Albedo, adimensional)	$â_1 = 0.64$, $â_2 = 0.2$
Window—to—wall ratio (WWR, %)	$W_1 = 20\%$, $W_2 = 40\%$
Climate zone	A4, B4, C4, D3, E1

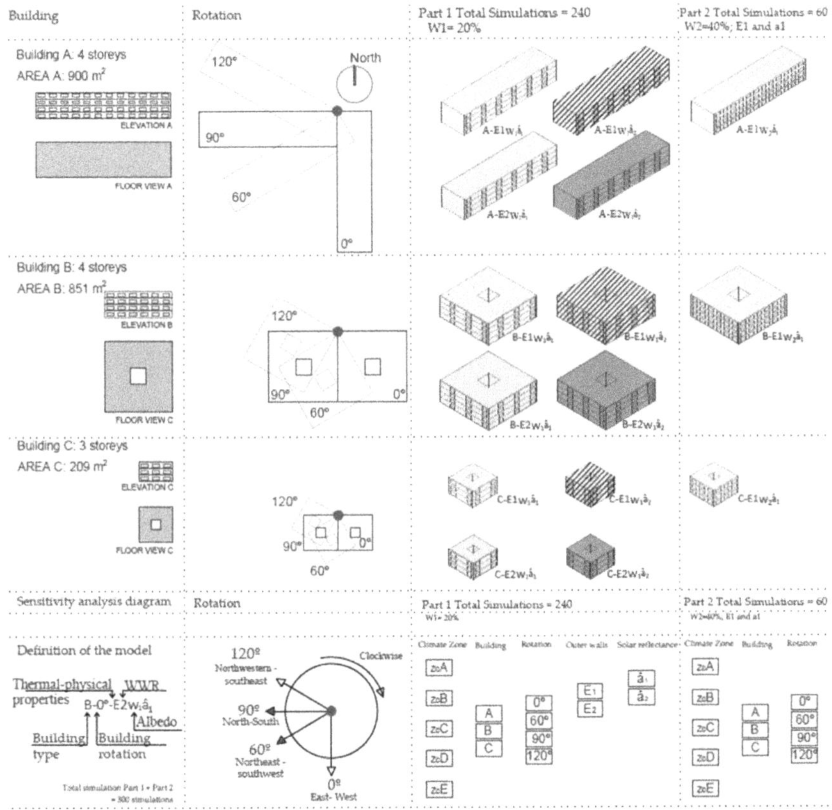

Figure 1. Simulated models according to the parameters and their nomenclature.

Table 3. Simulations schedules.

Hours	1–7	8	9–15	16–18	19	20–23	24
Cooling temperature (°C)							
June to September	27	—	—	25	25	25	27
Heating temperature (°C)							
January to May	17	20	20	20	20	20	17
October to December	17	20	20	20	20	20	17
Sensible heat from occupants (W/m²)							
Weekdays	2.15	0.54	0.54	1.08	1.08	1.08	2.15
Weekends	2.15	2.15	2.15	2.15	2.15	2.15	2.15
Latent heat from occupants (W/m²)							
Weekdays	1.36	0.34	0.34	0.68	0.68	0.68	1.36
Weekends	1.36	1.36	1.36	1.36	1.36	1.36	1.36
Lighting (W/m²)							
Everyday	0.44	1.32	1.32	1.32	2.2	4.4	2.2
Equipments (W/m²)							
Everyday	0.44	1.32	1.32	1.32	2.2	4.4	2.2

The first parameter considered is the orientation. The different orientations selected for the models are taken from the Spanish Building Technical Code (SBTC). In order to define the type of orientation of the different models, the 0° value corresponds to the East-West orientation whereas the 60° value corresponds to the Northeast-Southwest orientation. In addition, the value equal to 90° corresponds to the North-South orientation and the 120° value is related to the Northwest-Southeast orientation.

The second parameter considered is the overall heat transfer coefficient (U-value, $Wm^{-2}K^{-1}$), which determines the heat loss through the unit area of the envelope elements. It is common that regulations impose a maximum U-value to control the heat loss of buildings and thus ensure reduced energy consumption for heating and cooling. Since 68% of the current building stock in Spain was constructed before 1979 [51], two different thermal transmittance values are considered for the external walls, while keeping the U-values pertaining to the roof and to the windows as constant. The value of 1.58 $Wm^{-2}K^{-1}$ corresponds to those buildings constructed before the above mentioned regulation came into force, and the value of 0.35 $Wm^{-2}K^{-1}$ that corresponds to those built later. Table 4 describes all the transmittance values of walls, roof, and windows in order. Roof with a slope wasn't analyzed because its parametric modeling can cause numerical problems during the simulations [52–54].

Table 4. Constructive characteristics of the building envelope with their parametric variations.

Envelope Component	Layers	U-value ($Wm^{-2}K^{-1}$)
External walls E1	1) External Brickwork 105 mm Emissivity (ε): 0.9 Solar reflectance (â): $â_1 = 0.64$ $â_2 = 0.2$ Visible absorptance: 0.7 2) Standard Insulation 5.7 mm 3) Internal Brickwork 105 mm 4) Internal Plaster 13 mm	1.58 (not meeting the SBTC prescriptions)
External walls E2	1) Lightweight metal cladding e = 6 mm Emissivity: 0.52 Solar reflectance (â): $â_1 = 0.64$ $â_1 = 0.2$ 2) XPS Extruded Polystyrene 90 mm 3) Gypsum Plasterboard 13 mm	0.35 (meeting the SBTC prescriptions)
Windows (Constant Value)	1) Clear glazing 3 mm 2) Air gap 13 mm 3) Clear glazing 3 mm	1.96 (Table 2.3, meeting the SBTC prescriptions)
Roof (Constant Value)	1) Ceramic tile 20 mm 2) Mortar 20 mm 3) Waterproof coating 5 mm 4) Light—aggregate concrete 100 mm 5) Prefab hollow slab floor 200 mm 6) Gypsum plaster 15 mm 7) Air gap 100 mm 8) Rock wool 80 mm 9) Laminated gypsum board 15 mm	0.34 (meeting the SBTC prescriptions)

As far as solar reflectance is concerned, two different values have been considered: $â_1 = 0.64$ that corresponds to a light beige color, and the value $â_2 = 0.2$ that corresponds to a generic darker color.

The last parameter varied during the simulations is the Window-to-Wall Ratio (WWR). The WWR in the façade is a determining factor in terms of heat gains and losses through glazed surfaces. In order to define this variable, SBTC 2006 is taken as a reference. The percentages selected range from 21 to 30% and from 31 to 40%. It is considered that these values cover the usual spectrum of WWR in Spain [9].

2.2. Selection of Different Climate Zones

Research indicates that improving the insulation of buildings or saving the energy obtained from conventional sources must not be abandoned [11–13] Therefore, buildings with solar and cooling adaptation systems must be built since they are more effective in a climate like that of Spain. During the coldest winter month, i.e., January, 50% of Spanish cities could be heated with active and passive solar energy throughout the day. The remaining would need conventional heating only during the night. In an average winter month, such as November or March, 90% of the Spanish cities could be heated with solar contributions from passive and active systems. 80% of them would be worth heating only with passive solutions during the central hours of the day. In the hottest months, such as July in the interior zone or August on the coast areas, comfort conditions can be obtained in buildings by means of ventilation in the wettest areas, i.e., 26% of the cities. 56% of the other cities could be heated if buildings were capable of maintaining nighttime temperatures during the day. This is possible because the oscillation of temperatures ranges from 16 °C to 23 °C. Therefore, very few cities in Spain would need to consume energy in air conditioning under good conditions of mass and thermal inertia in buildings [16].

The climate zones established in the Basic Document of Energy Saving pertaining to the Technical Building Code (DB-H) [9] have been identified according to expected energy demand of buildings and gave rise to a total of 17 zones. Twelve of these zones are peninsular zones (established since 2006), namely those tagged as A3, A4, B3, B4, C1, C2, C3, C4, D1, D2, D3, and E1. In 2013, five specific climate zones were introduced corresponding to the Canary Islands (Alpha 1, 2, and 3, A2, and B2).

The nomenclature used defines the zones with a combination of its Winter Climate Severity (letter, α, A, B, C, D, E from the mildest to the coldest winter), and its Summer Climate Severity (number, 1–4, from the mildest to the hottest summer). The concept of climate severity is defined as the ratio of the energy demand of a building to that of the same building in a reference location.

In this work, five climate zones out of the 17 zones are selected. The maximum numerical value (numbers 1 to 4) is selected within each climate zone. It corresponds to the highest temperatures during the summer. Therefore, the selection of zones A4, B4, C4, D3, and E1 is considered. The representative cities for each zone are Almeria (latitude 36.85, longitude –2.38) within zone A4 and Seville (latitude 37.42, longitude –5.9) within zone B4. Caceres (latitude 39.47, longitude —6.33), Madrid (latitude 40.45, longitude –3.55), and Burgos (latitude 42.35, longitude –3.67) are considered within climate zones C4, D3, and E1 respectively (see Figure 2).

Figure 2. Spanish climate zones with the study ones highlighted.

Table 5 lists the main features of the selected climate zones. Their Heating Degree Days (HDD) and Cooling Degree Days (CDD) are first calculated on the basis of 20 and 25 °C respectively, while dry bulb temperature, relative humidity and global horizontal radiation values are presented for both the summer (from June to September) and the winter (January to March and October to December) periods as average seasonal values.

Table 5. Main characteristics of the different climates.

SBTC	Koppen-Geiger Classif. [a]	HDD.20 [b]	CDD.25 [b]	Dry Bulb Temperature (°C)	Relative Humidity (%)	Global Horizontal Radiation (Wh m^{-2} h^{-1})	Wind Speed (ms^{-1}) [c]
Winter							
A4	Cfa	1236	–	15.49	57.06	195.47	2.45
B4	Csa	1527	–	14.12	55.81	180.41	3.35
C4	Bsk	2088	–	11.56	56.38	142.36	3.10
D3	Cfa	2695	–	8.97	55.56	148.96	1.92
E1	Cfb	3234	–	6.69	56.44	143.37	4.65
Summer							
A4	Cfa	–	255	24.48	38.13	281.01	2.78
B4	Csa	–	219	24.46	35.50	280.89	3.56
C4	Bsk	–	256	22.82	38.00	264.01	2.52
D3	Cfa	–	172	22.82	35.38	263.48	1.72
E1	Cfb	–	32	17.99	37.88	223.46	4.54

[a] Koopen Classification according to Iberian Climate Atlas, by AEMET. [b] These data are extracted from Energy + Weather Data, with which the simulations have been accomplished as per SBTC Spanish Standard. [c] AEMET https://datosclima.es/Aemethistorico/Vientostad.php year 2018.

3. Results and Discussion

This section presents the results of the 300 simulations carried out for the different climate zones. The best and worst thermal models for each climate conditions are shown and commented, as well as the characteristics that mostly influence their energy demand. Finally, a comparison of the results according to the different climates and with respect to the PH standard compliance is reported.

3.1. Models Calibration and Graphical Presentation of the Results

As a measure of quality control of the simulations outcomes, the results of this research have been preliminary checked against the values established by the Spanish Institute of Energy Diversification and Saving (IDAE), which has determined the energy consumption and costs of the Spanish households thanks to in situ measurements of about 600 households in different climate zone of Spain. Such values can be considered as reference for both the heating and cooling demands in multi—dwelling units [55,56]. The Design Builder (DB) values presented in Figure 3 correspond to the average of the 300 models simulated. As it can be seen, the average values obtained with DB simulations are very close to those obtained from IDAE. The biggest difference is found when estimating the space heating demand in climate zone A4 (18% difference).

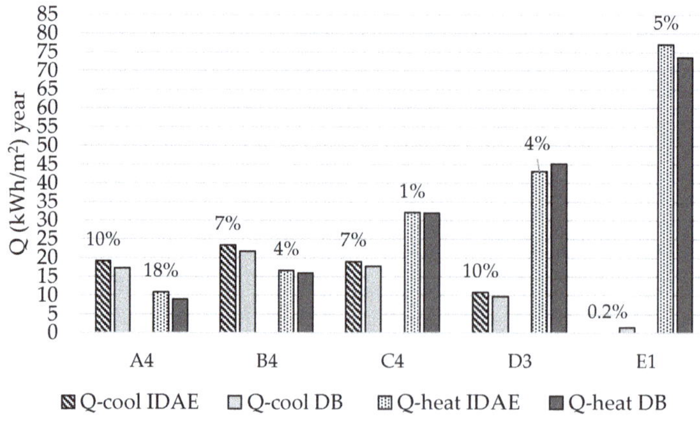

Figure 3. Comparison of total energy demand Q [kWh/m^2] with validated IDAE models.

Given the high number of simulations and the amount of resulting data, great attention has been paid also to the presentation of the results in a concise manner. As the purpose of this study is to examine the energy demand linked to the architectural design, the results are presented mainly through histograms and relying on the nomenclature and symbols defined in Section 2.1 that help to link the design strategy with the numerical results. The value equal to 15 kWhm^{-2}, and relevant to the PH standard energy demand threshold, has also been reported in the graphs in order to easily appreciate the fulfillment of the requirement.

Finally, the criterion chosen for defining the best design solution is that of the lowest energy demand for cooling (Q-cool), heating (Q-heat) and total (heating and cooling, Q). The opposite applies for identifying the worst design solutions.

3.2. Climate Zone A4

Figure 4 shows that the lowest heating energy demand is associated to model A (Q-heat = 0.84 kWhm^{-2}) when rotated 90° using envelope E2 (U = 0.35 Wm^{-2}K^{-1}), as well as a WWR equal to W1 (20%), dark colors â$_2$ (0.2), and a lower shape factor (0.26). On the other hand, the lowest cooling energy demand (Q-cool = 8.06 kWhm^{-2}) is associated to model B when rotated 0° and 90° using envelope E1 (U = 1.58 Wm^{-2}K^{-1}), as well as a WWR equal to W1 (20%), light colors â$_1$ (0.8), and a lower shape factor (0.26). Therefore, the optimal design parameters when considering both the heating and the cooling energy demands happen to occur when lower shape factors (0.26) and a lower WWR (20%) are taken into account.

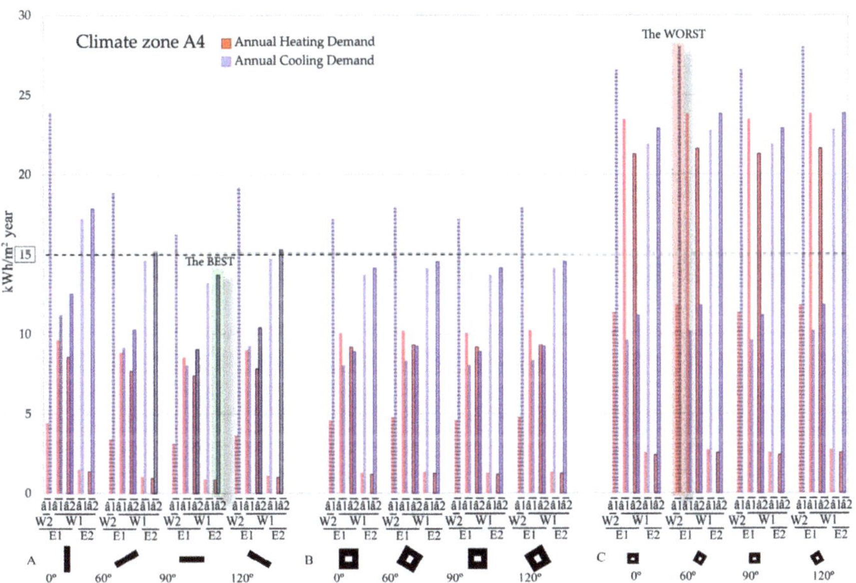

Figure 4. Heating and cooling energy demand in climate zone A4.

The worst model when considering the heating energy demand (Q-heat = 23.80 kWhm^{-2}) corresponds to model C when rotated 120° from the North using envelope E1 (U = 1.58 Wm^{-2}K^{-1}), as well as a WWR equal to W1 (20%), light colors â$_1$ (0.8), and a greater shape factor (0.48). As for the cooling energy demand, the highest value (Q-cool = 28.02 kWhm^{-2}) is associated to model C when rotated 60° and 120° from the North using envelope E1 (U = 1.58 Wm^{-2}K^{-1}), as well as a WWR equal to W2 (40%), light colors â$_1$ (0.8), and a greater shape factor (0.48). Due to this fact, the higher the

WWR (0.48), the greater the amount of solar radiation penetrating the interior of the building and therefore the more cooling energy demand needed [55].

On the other hand, when it comes to the appraisal of the total energy demand, the best model coincides with the best one pertaining to the lowest heating energy demand, except for the solar reflectance (â). In this particular case, the best model includes using light façade colors (â$_1$ = 0.8) with a value of the overall consumption equal to Q = 14.17 kWhm^{-2}. However, the worst model corresponds to model C (Q = 39.80 kWhm^{-2}) when rotated 60° and 120° from the North using envelope E1 (U = 1.58 Wm^{-2}K^{-1}), as well as a WWR equal to W2 (40%), light colors â$_1$ (0.8), and a greater shape factor (0.48). Therefore, the worst model with regard to the total energy demand coincides with the lowest cooling energy demand.

3.3. Climate Zone B4

Figure 5 shows the analysis of the results pertaining to climate zone B4. The best model within zone B4 regarding the lowest heating energy demand is associated to model A when rotated 90° using envelope E2 (U = 0.35 Wm^{-2}K^{-1}), as well as a WWR equal to W1 (20%), dark colors â$_2$ (0.2), and a lower shape factor (0.26). It corresponds to the same model and characteristics relevant to that of climate zone A4. In this particular case, the value of the heating energy demand equals 2.81 kWhm^{-2}. Such value doubles that of climate zone A4. As for the best model pertaining to the lowest cooling energy demand, it is associated to model B when rotated 0° and 90° using envelope E1 (U = 1.58 Wm^{-2}K^{-1}), as well as a WWR equal to W1(20%), light colors â1 (0.8), and a lower shape factor (0.26). It also coincides with the best model relating to climate zone A4 (Q-cool = 7.69 kWhm^{-2}), being slightly lower than the cooling energy demand relevant to climate zone A4.

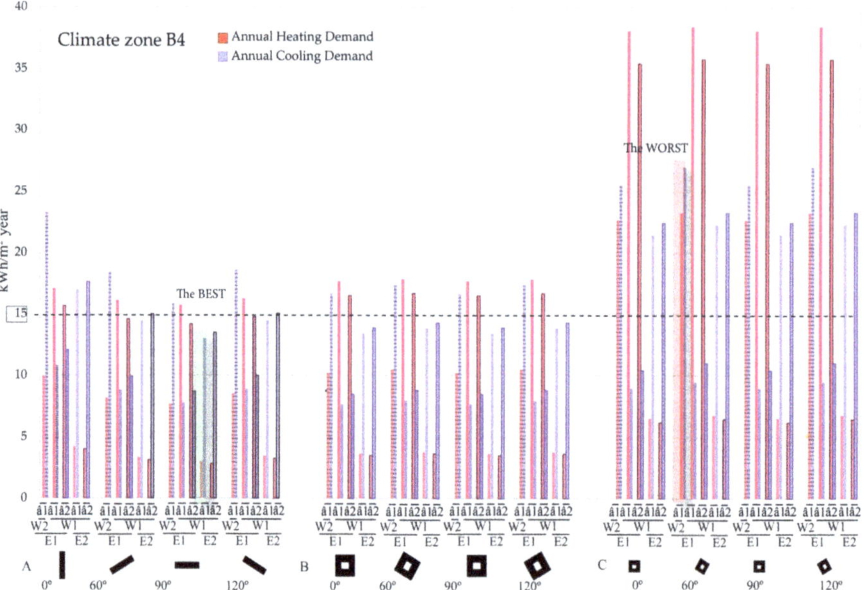

Figure 5. Heating and cooling energy demand in climate zone B4.

The worst model considering the highest heating energy demand (Q-heat = 38.44 kWhm^{-2}) corresponds to model C when rotated 60° using envelope E1 (U = 1.58 Wm^{-2}K^{-1}), as well as a WWR equal to W1 (20%), light colors â$_1$ (0.64), and a greater shape factor (0.48). In this particular case, the worst model is rotated 60° whereas the worst one is rotated 120° when considering climate zone A4. All the remaining parameters are equal to those pertaining to climate zone A4. The highest cooling energy demand (Q-cool = 26.97 kWhm^{-2}) is associated to model C when rotated 60° and 120° using envelope E1 (U = 1.58 Wm^{-2}K^{-1}), as well as a WWR equal to W2 (40%), light colors â1 (0.64), and a greater shape factor (0.48).

As for the total energy demand (Q = 16.05 kWhm^{-2}), the best model does not coincide with the value obtained in the case of the heating energy demand, since the envelope is different. Due to this reason, light colors are recommended. In addition, the worst model in terms of the total energy demand within this climate zone (Q = 50.26 kWhm^{-2}) corresponds to model C when rotated 60° and 120° using envelope E1 (U = 1.58 Wm^{-2}K^{-1}), as well as a WWR equal to W2 (40%), light colors â$_1$ (0.64), and a greater shape factor (0.48).

3.4. Climate Zone C4

Figure 6 shows the analysis of the results pertaining to climate zone C4. The best model within zone C4 is the same as those obtained within climate zones A4 and B4. The lowest heating energy demand (Q-heat = 11.16 kWhm^{-2}) is associated to model A when rotated 90° and using envelope E2 (U = 0.35 Wm^{-2}K^{-1}), as well as a WWR equal to W1 (20%), dark colors â$_2$ (0.2), and a lower shape factor (0.26). The lowest cooling energy demand (Q-cool = 4.3 kWhm^{-2}) is associated to model B when rotated 0° and 90° and using envelope E1 (U = 1.58 Wm^{-2}K^{-1}), as well as a WWR equal to W1 (20%), light colors â$_1$ (0.64), and a lower shape factor (0.26).

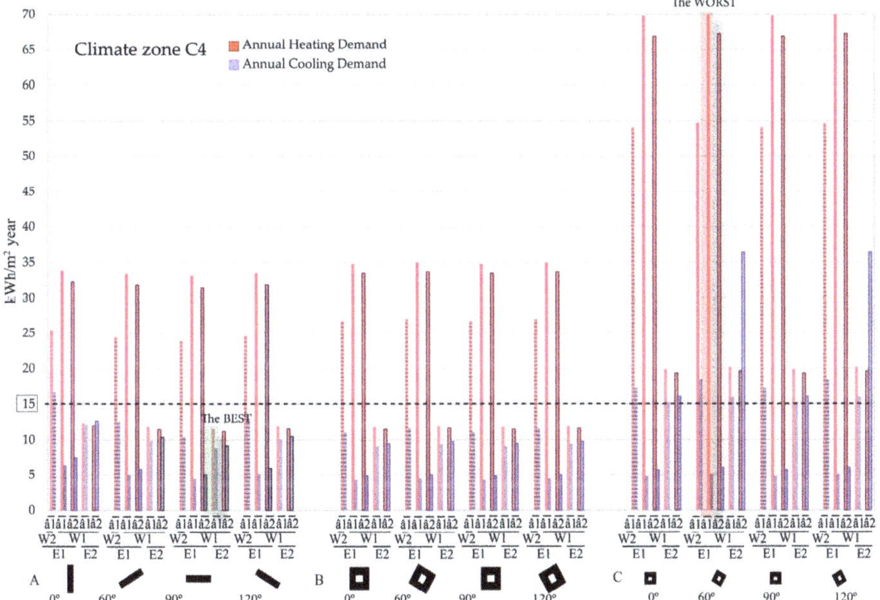

Figure 6. Heating and cooling energy demand in climate zone C4.

Considering the heating energy demand, the worst model coincides with that of climate zone B4 (Q-heat = 70.11 kWhm^{-2}). It corresponds to the highest heating energy demand associated to model C when rotated 60° and using envelope E1 (U = 1.58 Wm^{-2}K^{-1}), as well as a WWR equal to W1 (20%), light colors â$_1$ (0.64), and a greater shape factor (0.48) (i.e., C-60°-E1w$_1$â$_1$). As for the cooling energy demand, it coincides with the worst model pertaining to climate zones A4 and B4. The highest cooling energy demand (Q-cool = 18.43 kWhm^{-2}) is associated to model C when rotated 60° and 120° and using envelope E1 (U = 1.58 Wm^{-2}K^{-1}), as well as a WWR equal to W2 (40%), light colors â$_1$ (0.64), and a greater shape factor (0.48) (i.e., C-60° and 120°-E1w2â$_1$).

As for the total energy demand (Q), the optimal model coincides with those of climate zones A4 and B4. This configuration is given by model A (Q = 20.33 kWhm^{-2}) when rotated 90° and using envelope E2 (U = 0.35 Wm^{-2}K^{-1}), as well as a WWR equal to W1 (20%), light colors $\hat{a}_1$ (0.64), and a lower shape factor (0.26) (i.e., A-90°-E2w$_1\hat{a}_1$). However, the worst model in terms of the total energy demand within this climate zone C4 does not coincide with those of the remaining climate zones. The highest value of the total energy demand corresponds to model C (75.23 kWhm^{-2}) when rotated 60° and using envelope E1 (U = 1.58 Wm^{-2}K^{-1}), as well as a WWR equal to W1 (20%), light colors $\hat{a}_1$ (0.64), and a greater shape factor (0.48). (i.e., C-60°-E1w$_1\hat{a}_1$). Therefore, the worst model with regard to the total energy demand does not correspond to the worst one relevant to the cooling energy demand. This difference lies in the WWR. In fact, the worst model corresponds to a lower WWR (20%) whereas the worst model when considering climate zones A4 and B4 corresponds to a greater WWR (0.48).

3.5. Climate Zone D3

Figure 7 shows the analysis of the results pertaining to climate zone D3. The lowest heating energy demand within zone E1 is the same as those obtained within climate zones A4, B4, and C4. It corresponds to model A when rotated 90° and using envelope E2 (U = 0.35 Wm^{-2}K^{-1}), as well as a WWR equal to W1 (20%), dark colors $\hat{a}_1$ (0.64), and a lower shape factor (0.26). This corresponds to the model A-90°-E2w1$\hat{a}_2$ with a value of the heating energy demand equal to 20.29 kWhm^{-2}. However, the lowest cooling energy demand within zone D3 corresponds to model B (Q-cool=4.29 kWhm^{-2}) when rotated 0° and 90° and using envelope E1 (U = 1.58 Wm^{-2}K^{-1}), as well as a WWR equal to W1 (20%), light colors $\hat{a}$1 (0.64), and a lower shape factor (0.26), i.e., B-0° and 90°-E1w1$\hat{a}$1.

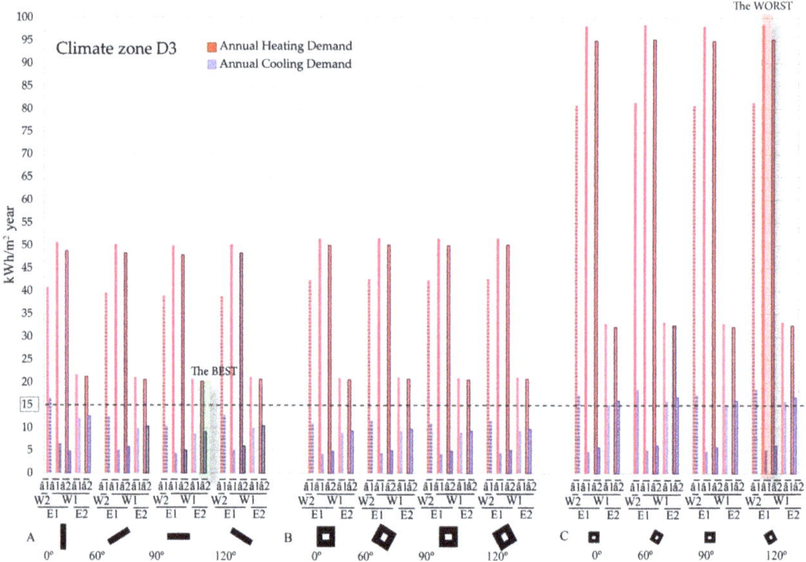

Figure 7. Heating and cooling energy demand in climate zone D3.

The highest heating energy demand within zone D3 is the same as that of climate zone A4. This model corresponds to a greater value of the heating energy demand associated to model C (Q-heat=98.51 kWhm^{-2}) when rotated 120° and using envelope E1 (U = 1.58 Wm^{-2}K^{-1}), as well as a WWR equal to W1 (20%), light colors $\hat{a}_1$ (0.64), and a greater shape factor (0.48) (i.e., C-120°-E1w1$\hat{a}_1$). The highest cooling energy demand (Q-cool = 18.44 kWhm^{-2}) coincides with those of climate zones A4, B4, and C4. In this case it corresponds to a greater value of the cooling energy demand associated to model C when rotated 60° and 120° and using envelope E1 (U = 1.58 Wm^{-2}K^{-1}), as well as a WWR equal to W2 (40%), light colors â1 (0.64), and a greater shape factor (0.48) (i.e., C-60° and 120°-E1w2â$_1$).

With regard to the total energy demand (Q), the best model is different from those of the climate zones previously considered. In this particular case, such best model corresponds to model A (Q = 29.41 kWhm^{-2}) when rotated 90° and using envelope E2 (U = 0.35 Wm^{-2}K^{-1}), as well as a WWR equal to W1 (20%), dark colors â$_2$ (0.2), and a lower shape factor (0.26). For the first time the use of dark colors results in a reduction of the overall consumption. On the other hand, the worst model does not coincide with the one defined for zone C4. In this particular case, it corresponds to model C (Q=103.63 kWhm^{-2}) when rotated 120° and using envelope E1 (U = 1.58 Wm^{-2}K^{-1}), as well as a WWR equal to W1 (20%), light colors â$_1$ (0.64), and a greater shape factor (0.48) (C-120°-E1w1â$_1$). The orientation in this case is 120° whereas that of zone C4 is 60°.

3.6. Climate Zone E1

Figure 8 shows the analysis of the results pertaining to climate zone E1. The lowest heating energy demand within zone E1 is the same as those obtained within all the climate zones previously considered. It corresponds to model A-90°-E2w1â$_2$ with a value of heating energy demand equal to 30.14 kWhm^{-2}. However, the lowest cooling energy demand within zone E1 corresponds to model C (Q-cool = 0.14 kWhm^{-2}) when rotated 90° and using envelope E1 (U = 1.58 Wm^{-2}K^{-1}), as well as a WWR equal to W1 (20%), light colors â$_1$ (0.64), and a greater shape factor (0.48).

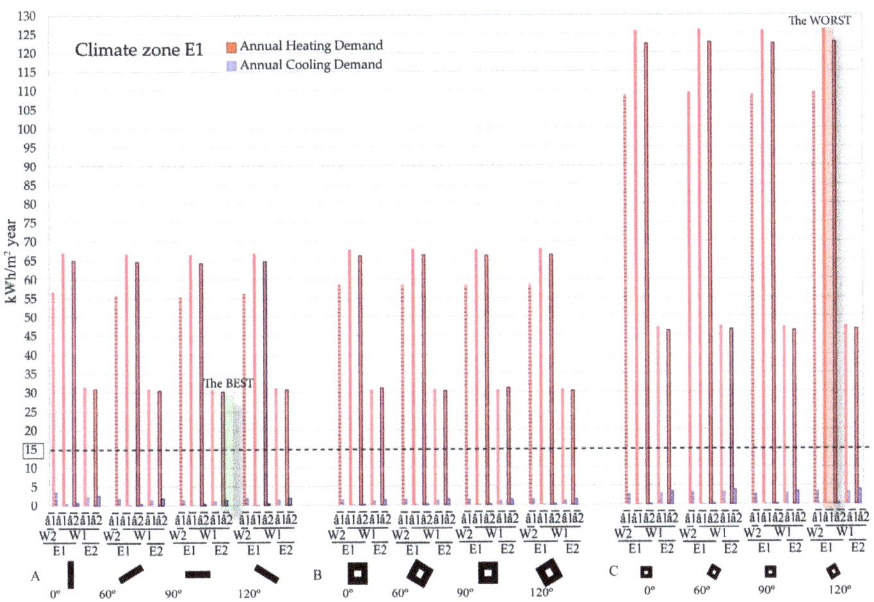

Figure 8. Heating and cooling energy demand in climate zone E1.

The highest heating energy demand within zone E1 is the same as that of climate zone D3. This worst model corresponds to the highest heating energy demand associated to model C (Q-heat=126.04 kWhm^{-2}) when rotated 120° using envelope E1 (U = 1.58 Wm^{-2}K^{-1}), as well as a WWR equal to W1 (20%), light colors â$_1$ (0.64), and a greater shape factor (0.48). The worst model in terms of the cooling energy demand (Q-cool = 4.01 kWhm^{-2}) is different from the rest of the zones analyzed previously. In this case it corresponds to the highest cooling energy demand associated to model C when rotated 120° and using envelope E2 (U = 0.35 Wm^{-2}K^{-1}), as well as a WWR equal to W1(20%), dark colors â$_2$ (0.8), and a greater shape factor (0.48). The parameter varied is the transmittance of the wall E2 ($U = 0.35$ Wm^{-2}K^{-1}). In the previous climate zones, it corresponds to the value E1 ($U = 1.58$ Wm^{-2}K^{-1}).

The lowest total energy demand within zone E1 is the same as that of zone D3. The best model corresponds to model A ($Q = 31.67$ kWhm^{-2}) when rotated 90° and using envelope E2 ($U = 0.35$ Wm^{-2}K^{-1}), as well as a WWR equal to W1 (20%), dark colors $â_2$ (0.2), and a lower shape factor (0.26). On the other hand, the highest total energy demand is that of zone D3. In this particular case it corresponds to the highest total energy demand associated to model C ($Q = 126.19$ kWhm^{-2}) when rotated 120° using envelope E1 ($U = 1.58$ Wm^{-2}K^{-1}), as well as a WWR equal to W1 (20%), light colors $â_1$ (0.64), and a greater shape factor (0.46). Both models are similar to that of climate zone D3.

3.7. Climates Cross-Comparison

Table 6 shows the comparison between the optimal models and the worst ones with respect to the different climate zones. The results show how building A, which corresponds to a linear block, is the optimal shape with regard to the heating energy demand and the total energy demand. They also indicate that model C, which is the one with the greatest shape factor (0.48), presents the highest energy demand values. Heating in Spain is the main responsible for energy consumption. In colder climates such as E1, the amount of heat loss through the envelope is greater than the amount of heat that can be gained when increasing the external surface receiving solar radiation. Therefore, a proportionality is established between the increment of the shape factor (i.e., with a lower index of compactness) and the increment of the energy needed for heating [57]. Accordingly, this research work shows that the optimal model is that with the lowest shape factor.

Table 6. Summary of the best and worst configurations for each climate.

	Best Models			Worst Models		
Climate zone	Q_heat	Q_cool	Q	Q_heat	Q_cool	Q
A4	A-90°- E2w1$â_2$	B-0/90°- E1w1$â_1$	A-90°- E2w1$â_1$	C-120°- E1w1$â_1$	C-60°/120°- E1w2$â_1$	C-60° and 120°-E1w2$â_1$
B4	A-90°- E2w1$â_2$	B-0/90°- E1w1$â_1$	A-90°- E2w1$â_1$	C-60°- E1w1$â_1$	C-60°/120°- E1w2$â_1$	C-60° and 120°-E1w2$â_1$
C4	A-90°- E2w1$â_2$	B-0/90°- E1w1$â_1$	A-90°- E2w1$â_1$	C-60°- E1w1$â_1$	C-60°/120°- E1w2$â_1$	C-60°- E1w1$â_1$
D3	A-90°- E2w1$â_2$	B-0/90°- E1w1$â_1$	A-90°- E2w1$â_2$	C-120°- E1w1$â_1$	C-60°/120°- E1w2$â_1$	C-120°- E1w1$â_1$
E1	A-90°- E2w1$â_2$	C-0/90°- E1w1$â_1$	A-90°- E2w1$â_2$	C-120°- E1w1$â_1$	C-120°- E2w2$â_1$	C-120°- E1w1$â_1$

As for the orientation, the North-South orientation (0° and 90°) of the main façades is the best one. The design of buildings oriented to the North-South axis is recommended, with the largest façade oriented to the south. Orientation is one of the parameters that has the most impact on the design of buildings. Most books, guides or manuals on passive solar techniques recommend a southerly orientation, although consensus is based on the fact that the best option is an orientation of 20–30° to the south [58]. This has to do with the fact that the southern side of a building receives the maximum amount of solar radiation whereas the northern side receives the minimum amount of solar radiation (i.e., just its diffuse component).

Accordingly, in the current research, the optimal orientation found in the five climate zones for models A, B and C is 90° (North-South), the best model being A while the worst being C.

As for the thermal transmittance of the external walls, the best configuration is that complying with the SBTC (0.35 Wm^{-2}K^{-1}) and showing the lowest WWR (20%). In terms of solar reflectance, it is found that in climate zones A4, B4, and C4 the use of light colors is better, whereas in climate zones D3 and E1 the best option is using darker colors.

On the other hand, the worst configuration is found for the C-60° and 120° from the North model, with walls U-value not complying with the SBTC prescriptions (1.58 Wm^{-2}K^{-1}). Regarding the WWR, in zones A4 and B4 the use of larger WWR's is unfavorable, while the use of lower WWR is unfavorable when considering zones C4, D3, and E1. With respect to solar reflectance, it is pointed out that the worst case is given by the use of light colors in all climates.

3.8. Sensitivity Analysis of the Most Important Parameters

This section comments the results of a sensitivity analysis carried out on the parameters architects and design professionals should particularly take care of as they mostly affect the energy needs of the buildings: WWR, solar reflectance and U-values of the external walls.

Figure 9 analyzes the variations of these parameters according to all the climate zones, with positive values expressing an increase and negative values indicating reductions.

The graphic is divided into three parts; each of them corresponds to a specific variation in the architectural design. The first variation corresponds to the reduction of the WWR from 40% to 20%, and is depicted in the top row.

The second one is the variation from light to darker color when the thermal transmittance of the envelope meets SBTC prescriptions. The base model has a value of $\hat{a}_1$ equal to 0.64, while the final model has a value $\hat{a}_2$ equal to 0.2. The third variation is that from a light color to a darker one when the thermal transmittance does not comply with SBTC.

Figure 9 shows that, when the WWR is reduced (top row), the heating energy demand gets increased. The increments obtained in the different climates are of 6.91 kWhm^{-2} for climate zone A4, 8.29 kWhm^{-2} for climate zone B4, 7.73 kWhm^{-2} for zone C4, 8.35 kWhm^{-2} for zone D3 and 8.02 kWhm^{-2} for climate zone E1.

The second variation (i.e., middle row) corresponds to the shift from light to dark colors when the thermal transmittance of the envelope does not meet the SBTC requirements. It is found that in such particular case the cooling energy demand increases as following: in climate A4 of 0.78 kWhm^{-2}, in climate B4 of 0.73 kWhm^{-2}, in climate C4 the increase is of 0.47 kWhm^{-2}, while for climates D3 and E1 they amount to 0.46 kWhm^{-2} and 0.14 kWhm^{-2} respectively. On the other hand, it can be noted that the heating energy demand gets reduced in a range of 1.3 kWhm^{-2} for climate zone A4 to 1.88 kWhm^{-2} for climate zone E1.

Finally, the third variation corresponds to a shift from light to dark colors when the thermal transmittance does not comply with the SBTC requirements. It is found that the heating demand gets reduced of the following amount: 0.59 kWhm^{-2} in climate zone A4, 0.58 kWhm^{-2} in climate zone B4, 0.53 kWhm^{-2} in climate zone C4, 0.52 kWhm^{-2} in climate zone D3 and 0.31 kWhm^{-2} in climate zone E1. On the other hand, the cooling energy demand increases in a range of 0.12 kWhm^{-2} for climate zone A4 to 0.45kWhm^{-2} for climate zone E1.

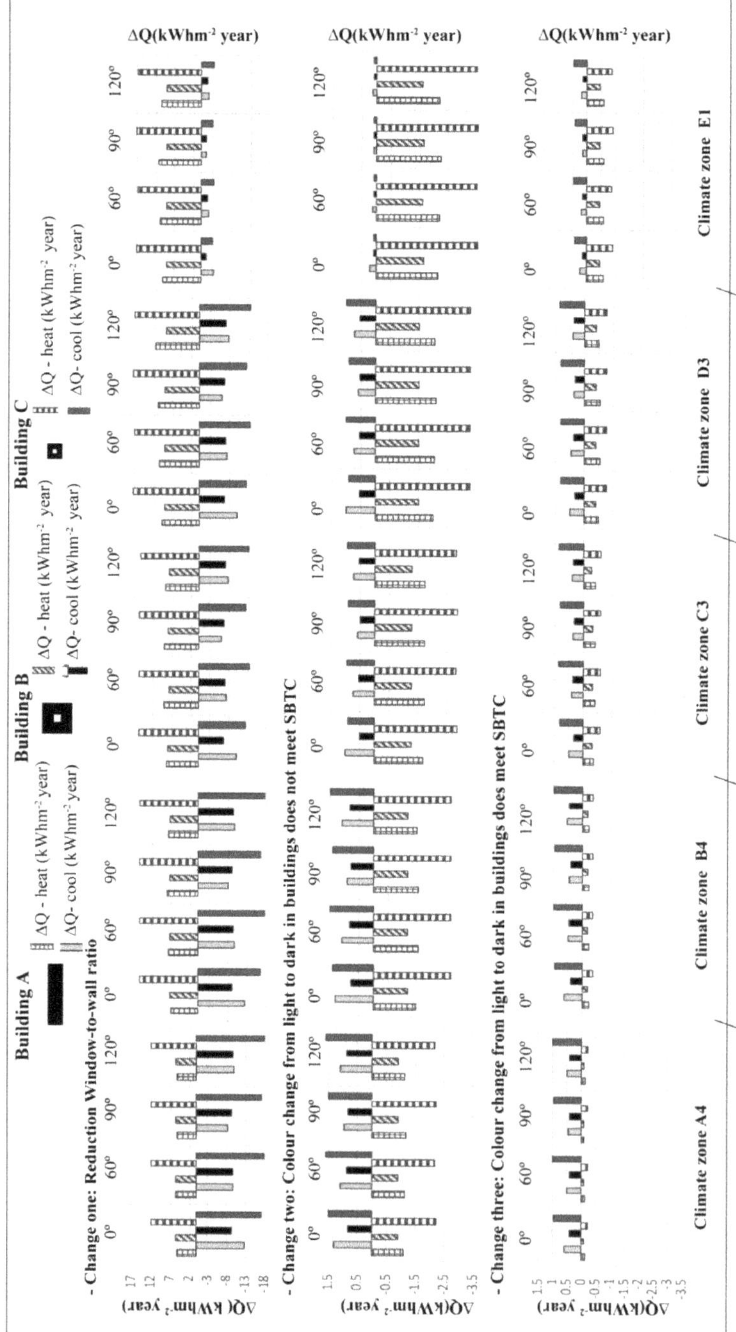

Figure 9. Sensitivity analysis of the energy demand differences when changing WWR (top row), solar reflectance (middle row) and U-values (bottom row).

4. Compliance with the PH Standard

This section deals with the compliance of the simulated models with the PH threshold of 15kWhm^{-2} for heating and cooling purposes. Red values shown in Figure 10 correspond to the heating values that are lower than the PH standard, while the blue values correspond to the cooling values meeting the PH standard.

Building	Type model	C.Z. A4		C.Z. B4		C.Z. C4		C.Z. D3		C.Z. E1	
		Q-heat	Q-cool	Q-heat	Q-cool	Q-heat	Q-cool	Q-heat	Q-cool	Q-heat	Q-cool
Rotation 0	A-E1w2â1	4.42	23.90	9.95	23.30	25.40	16.61	40.80	16.60	56.77	3.78
	A-E1w1â1	9.64	11.21	17.06	10.78	33.90	6.46	50.66	6.47	67.01	0.46
	A-E1w1â2	8.54	12.55	15.64	12.10	32.24	7.45	48.69	7.47	64.87	0.67
	A-E2w1â1	1.51	17.25	4.21	16.97	12.28	12.05	21.74	12.01	31.52	2.47
	A-E2w1â2	1.38	17.87	3.99	17.60	11.91	12.58	21.26	12.53	30.94	2.70
Rotation 60	A-E1w2â1	3.40	18.86	8.18	18.43	24.42	12.50	39.62	12.49	55.82	2.17
	A-E1w1â1	8.83	9.16	16.09	8.82	33.46	5.06	50.22	5.07	66.69	0.24
	A-E1w1â2	7.68	10.26	14.58	9.91	31.77	5.80	48.19	5.80	64.53	0.36
	A-E2w1â1	1.08	14.65	3.35	14.46	11.81	9.92	21.16	9.88	31.02	1.73
	A-E2w1â2	0.97	15.18	3.13	14.99	11.43	10.36	20.64	10.32	30.43	1.90
Rotation 90	A-E1w2â1	3.14	16.25	7.65	15.84	23.91	10.45	39.00	10.44	55.35	1.52
	A-E1w1â1	8.54	8.08	15.71	7.76	33.14	4.41	49.91	4.38	66.43	0.15
	A-E1w1â2	7.39	9.03	14.18	8.68	31.41	5.02	47.84	4.98	64.22	0.24
	A-E2w1â1	0.94	13.24	3.02	13.03	11.54	8.79	20.83	8.73	30.74	1.39
	A-E2w1â2	0.84	13.70	2.81	13.49	11.16	9.17	20.29	9.12	30.14	1.52
Rotation 120	A-E1w2â1	3.61	19.19	8.52	18.64	24.62	12.82	38.88	12.79	56.20	2.29
	A-E1w1â1	8.98	9.27	16.27	8.88	33.53	5.18	50.31	5.15	66.82	0.26
	A-E1w1â2	7.84	10.38	14.77	9.97	31.82	5.92	48.27	5.90	64.65	0.38
	A-E2w1â1	1.14	14.77	3.48	14.49	11.89	10.04	21.23	9.97	31.19	1.78
	A-E2w1â2	1.03	15.29	3.27	15.03	11.51	10.48	20.75	10.41	30.61	1.95
Rotation 0	A-E1w2â1	4.59	17.21	10.20	16.66	26.71	11.04	42.37	11.05	58.52	1.59
	A-E1w1â1	10.05	8.06	17.67	7.69	34.78	4.30	51.52	4.29	67.75	0.14
	A-E1w1â2	9.16	8.88	16.51	8.48	33.51	4.82	50.03	4.82	66.13	0.21
	A-E2w1â1	1.30	13.72	3.67	13.46	11.76	9.11	21.04	9.07	30.59	1.42
	A-E2w1â2	1.21	14.14	3.50	13.89	11.47	9.45	20.65	9.39	30.14	1.55
Rotation 60	A-E1w2â1	4.77	17.97	10.47	17.40	26.96	11.64	42.64	11.65	58.80	1.79
	A-E1w1â1	10.19	8.35	17.83	7.98	34.93	4.48	51.66	4.48	67.89	0.16
	A-E1w1â2	9.30	9.21	16.68	8.81	33.67	5.02	50.17	5.04	66.29	0.24
	A-E2w1â1	1.37	14.12	3.79	13.86	11.89	9.44	21.18	9.40	30.73	1.53
	A-E2w1â2	1.27	14.56	3.63	14.30	11.60	9.79	20.79	9.75	30.28	1.66
Rotation 90	A-E1w2â1	4.59	17.21	10.20	16.66	26.71	11.04	42.37	11.05	58.52	1.59
	A-E1w1â1	10.05	8.06	17.67	7.69	34.78	4.30	51.52	4.29	67.75	0.14
	A-E1w1â2	9.16	8.88	16.51	8.48	33.51	4.82	50.03	4.82	66.13	0.21
	A-E2w1â1	1.30	13.72	3.67	13.46	11.76	9.11	21.04	9.07	30.59	1.42
	A-E2w1â2	1.21	14.14	3.50	13.89	11.47	9.45	20.65	9.39	30.14	1.55
Rotation 120	A-E1w2â1	4.79	17.97	10.48	17.38	26.98	11.65	42.72	11.66	58.89	1.79
	A-E1w1â1	10.21	8.36	17.83	7.98	34.93	4.49	51.69	4.48	67.93	0.16
	A-E1w1â2	9.31	9.22	16.69	8.81	33.67	5.03	50.21	5.05	66.33	0.24
	A-E2w1â1	1.37	14.13	3.80	13.85	11.87	9.44	21.22	9.40	30.79	1.53
	A-E2w1â2	1.28	14.57	3.63	14.30	11.61	9.81	20.83	9.76	30.34	1.66
Rotation 0	A-E1w2â1	11.34	26.61	22.70	25.56	54.10	17.28	80.89	17.31	108.66	3.04
	A-E1w1â1	23.47	9.63	38.10	8.99	69.77	4.81	98.20	4.78	125.71	0.12
	A-E1w1â2	21.27	11.16	35.43	10.42	66.93	5.71	94.93	5.70	122.24	0.20
	A-E2w1â1	2.59	21.93	6.53	21.48	19.89	15.23	32.98	15.17	47.04	3.26
	A-E2w1â2	2.39	22.90	6.19	22.46	19.31	16.06	32.21	15.99	46.14	3.65
Rotation 60	A-E1w2â1	11.78	28.02	23.29	26.97	54.68	18.43	81.46	18.44	109.28	3.49
	A-E1w1â1	23.79	10.17	38.44	9.50	70.11	5.12	98.50	5.10	126.03	0.15
	A-E1w1â2	21.61	11.77	35.78	11.02	67.29	6.07	95.24	6.09	122.58	0.25
	A-E2w1â1	2.75	22.78	6.79	22.32	20.17	15.95	33.28	15.88	47.35	3.57
	A-E2w1â2	2.54	23.80	6.45	23.31	19.59	16.80	32.52	16.73	46.47	4.00

Figure 10. *Cont.*

Rotation 90	A-E1w2â1	11.34	26.61	22.70	25.56	54.10	17.28	80.89	17.31	108.66	3.04
	A-E1w1â1	23.47	9.63	38.10	8.99	69.77	4.81	98.20	4.78	125.71	0.12
	A-E1w1â2	21.27	11.16	35.43	10.42	66.93	5.71	94.93	5.70	122.24	0.20
	A-E2w1â1	2.59	21.93	6.53	21.48	19.89	15.23	32.98	15.17	47.04	3.26
	A-E2w1â2	2.39	22.90	6.19	22.46	19.31	16.06	32.21	15.99	46.14	3.65
Rotation 120	A-E1w2â1	11.78	28.02	23.29	26.97	54.68	18.43	81.46	18.44	109.28	3.49
	A-E1w1â1	23.80	10.21	38.42	9.52	70.06	5.14	98.51	5.12	126.04	0.15
	A-E1w1â2	21.61	11.80	35.77	11.04	67.24	6.11	95.27	6.11	122.61	0.25
	A-E2w1â1	2.76	22.82	6.79	22.33	20.15	15.98	33.33	15.92	47.41	3.59
	A-E2w1â2	2.55	23.83	6.45	23.35	19.57	16.85	32.56	16.77	46.53	4.02

Figure 10. Meeting the PH standard requirements.

The first conclusion that can be drawn is that buildings located in climate zones C4, D3, and E1 deserve more detailed studies since their design according to basic passive design measures does not allow meet the PH standard requirements. Looking closer, models A and B that present a shape factor of 0.26 generally meet the PH criterion when considering climate zones A4 and B4. On the contrary, model C does not meet the standards in such climates.

When the heating energy demand column is analyzed, it can be noted that in climate zones A4 and B4 most of the demand is below 15 kWhm^{-2}. On the other hand, as long as climate zones become more severe during the winter (i.e., climate zones D3 and E1), the heating energy demand is never below 15 kWhm^{-2}. If the shape factor is considered, models A and B (i.e., those models with a shape factor of 0.26) do meet the PH standard, whereas the heating demand pertaining to model C is seldom below 15 kWhm^{-2}.

As for the cooling energy demand column, the opposite results are obtained. It is observed that in climate zones E1 and D3 most of the cooling demand is below 15 kWhm^{-2}. On the other hand, as long as climate zones become more severe during the summer (i.e., climate zones A4 and B4), only few models present demand values below 15 kWhm^{-2}. As far as the shape factor is concerned, the same behavior described above is found. Models A and B do meet the PH standard, whereas if considering model C the cooling demand is seldom below 15 kWhm^{-2}. According to Figure 10, in climate zones such as those of Spain, heating must be the main focus with regard to the design of buildings in order to comply with the PH standard.

5. Conclusions

This research work is intended to support building designers and professionals during the early design stages of residential buildings located in Spain by providing an energy analysis of different architectural strategies typically implemented in passive design.

Three building types representative of the multi-dwelling units (MDUs) stock are simulated through a parametric analysis in Design Builder. For each typology, parameters such as orientation, Window-to-Wall Ratio (WWR), shape factor, outer walls construction, color of the envelope, and climate zones are evaluated, resulting in 300 simulations. Such parameters have been graphed and related with the building design, so that the energy demand for space heating and cooling can be directly related with the specific design parameters, and the communication gap between simulation outputs and architects' understanding of them closed.

Regarding the heating demand, the best model turns out to be A-90°-E2w1â2 for all climates. It is defined by using the thermal transmittance that meets the SCTB (U = 0.35 Wm^{-2}K^{-1}), as well as a low WWR (20%) along with dark colors (â2 = 0.2). Regarding the worst model, it is found to be the C-60° and 120° from the North when using a thermal transmittance that does not comply with the SCTB (U = 1.58 Wm^{-2}K^{-1}), as well as the lowest percentage of WWR (20%) and light colors (â1 = 0.64).

Regarding the cooling demand, the best model is not the same in all climate zones. In zones A4, B4, C4, and D3, it is B-0° oriented 90° from the North when using a thermal transmittance that does not comply with the SCTB as well as the lowest percentage of WWR (20%) and light colors (E1w1â1).

As for the E1 climate zone, the best configuration is found when using external walls whose thermal transmittance does not comply with the SCTB (U = 1.58 Wm^{-2}K^{-1}), having a low value of the WWR parameter (20%) and light colors for finishing ($\hat{a}_1$ = 0.64). Regarding the worst model, it always turns out to be the same one in all climate zones, i.e. C—60° and 120° from the North-E1w2$\hat{a}_1$. In such cases the thermal transmittance does not comply with the SCTB (U = 1.58 Wm^{-2}K^{-1}), the greatest percentage of WWR (40%) is used and the colors considered are light ($\hat{a}_1$ = 0.64).

Regarding the total energy demand, the best model in climate zones A4, B4, and C4 is A-90°-E2w1$\hat{a}_1$. It is defined using the thermal transmittance that meets the SCTB (U= 0.35 Wm^{-2}K^{-1}), as well as the lowest percentage of WWR (W1 = 20%) when using light colors ($\hat{a}_1$ = 0.64). On the contrary, the best model in climate zones D3 and E1 is A-90°-E2w1$\hat{a}_2$. It is defined using the thermal transmittance that meets the SCTB (U = 0.35 Wm^{-2}K^{-1}), as well as the lowest percentage of WWR (W1 = 20%) when using dark colors ($\hat{a}_2$ = 0.2). Regarding the worst model in climate zones A4 and B4, it happens to be the C-60° and 120° ones when using the thermal transmittance of the wall that does not comply with the SCTB (U = 1.58 Wm^{-2}K^{-1}), as well as the lowest percentage of WWR (20%) when using light colors ($\hat{a}_1$ = 0.64). Regarding climate zones C4, the worst model is C-60°-E1w1$\hat{a}_1$. It is defined using the thermal transmittance that does not comply with the SCTB (U = 1.58 Wm^{-2}K^{-1}), as well as the lowest percentage of WWR (W1 = 20%) when using light colors ($\hat{a}_1$ = 0.64). As for climate zones D3 and E1, the worst model is C-120°-E1w1$\hat{a}_1$. It is defined using the thermal transmittance that does not comply with the SCTB (U = 1.58 Wm^{-2}K^{-1}), as well as the lowest percentage of WWR (W1=20%) when using light colors ($\hat{a}_1$ = 0.64).

Most parameters produce contradictory effects when comparing the heating and cooling demands. As for the shape factor, if considering square building the heating energy demand reaches its minimum value when the façades are oriented directly to the four cardinal points. In a rectangular building, the heating demand is reduced as the façade facing east is smaller. The reduction of the WWR causes the heating demand to decrease and the cooling demand to increase; this is beneficial for climate zone E1 but harmful for A4. In addition, the reduction in the WWR when considering A4 climates helps the users to reduce the cooling energy demand while the same action in cooler areas worsens the heating energy demand.

Regarding the reflectance of the materials, in climate zone A4 the use of reflective and selective cold materials is beneficial. When considering E1 (U − 1.58 Wm^{-2}K^{-1}), the use of absorptive materials is definitely recommended. In cooler zones such as E1, it is instead recommended to use darker colors while in warmer zones such as A4 it is recommended to use lighter colors ($\hat{a}_1$ = 0.64).

Finally, those buildings located in climate zones C4, D3, and E1 require more detailed analysis since their design does not meet the PH standard according to the basic passive design measures discussed. It was demonstrated the importance to applying the passive house principles in early design phases in order to reduce energy consumptions. The adoption of useful tool like dynamic simulations software can help architects, designs, builders and stakeholders in the decision—making process to recognize that the PH model is one of the most effective ways of reducing carbon emissions.

More efficient designs can be obtained by using appropriate passive strategies which will be accomplished by building designers committed and aware of the current situation in order to comply with the requirements of European energy policies. Although considering the users' habits is still a pending issue, using passive strategies is definitely a breakthrough in the design of efficient buildings.

Author Contributions: Conceptualization, M.-M.F.-A., R.-A.G.-L.; methodology, R.-A.G.-L.; software, M.-M.F.-A., V.C., F.N.; validation, M.-M.F.-A., R.-A.G.-L.; formal analysis, M.-M.F.-A., V.C., F.N.; investigation, M.-M.F.-A.; resources, R.-A.G.-L.; data curation, J.M.d.R.; writing—original draft preparation, M.-M.F.-A., F.N.; writing—review and editing, M.-M.F.-A., V.C., F.N.; visualization, R.-A.G.-L.; supervision, V.C., F.N., R.-A.G.-L.; project administration, R.-A.G.-L., F.N.; funding acquisition, R.-A.G.-L., F.N.

Funding: The authors wish to thank Fundación Universitaria CEU San Pablo for the predoctoral scholarship granted to co-author Maria-Mar Fernandez-Antolin within its FPI Program.

Acknowledgments: Thanks are due to Arie Group from the Institute of Technology within Universidad CEU San Pablo and the laboratory within Universidad CEU San Pablo because of the Design Builder software licenses provided. Finally, the authors wish to thank both CEINDO CEU and Banco Santander for the international mobility scholarship granted to co-author Maria-Mar Fernandez-Antolin in order to develop an external research stay at Department of Civil Engineering and Architecture at the University of Catania.

Conflicts of Interest: The author(s) declared no potential conflicts of interest with respect to the research, authorship, and/or publication of this article.

References

1. Tettey, U.Y.A.; Dodoo, A.; Gustavsson, L. Design strategies and measures to minimise operation energy use for passive houses under different climate scenarios. *Energy Effic.* **2019**, *12*, 299–313. [CrossRef]
2. EC Directive. 2010/31/EU of the European Parliament and of the Council of 19 May 2010 on the Energy Performance of Buildings (Recast). 2010. Available online: http://eur---lex.europa.eu/legal---content/EN/TXT/?uri=CELEX:32010L0031 (accessed on 7 May 2019).
3. EC Directive. *2012/27/EU of the European Parliament and of the Council of 25 October 2012 on Energy Efficiency, Amending Directives 2009/125/EC and 2010/30/EU and Repealing Directives 2004/8/EC and 2006/32/EC Text with EEA Relevance*; EC Directive: Bruxels, Belgium, 2012.
4. EPBD recast. Directive 2018/844/UE of the European Parliament and of Council, 30 May 2018 on the energy performance of buildings (recast). *Off. J. Eur. Union* **2018**, *156*, 75–91.
5. Nejat, P.; Jomehzadeh, F.; Taheri, M.M.; Gohari, M.; Majid, M.Z.A. A global review of energy consumption, CO2 emissions and policy in the residential sector (with an overview of the top ten CO2 emitting countries). *Renew. Sustain. Energy Rev.* **2015**, *43*, 843–862. [CrossRef]
6. Institute for Energy Diversification and Savings (IDAE). *Spanish Energy Saving and Efficiency Plan 2014–2020*; IDAE: Madrid, Spain, 2014; p. 91.
7. Institute for Energy Diversification and Savings (IDAE). *Spanish Energy Saving and Efficiency Plan 2011–2020*; IDAE: Madrid, Spain, 2010; p. 195.
8. Technical building Code. *Basic Document HE. "Energy Saving", Spanish Technical Building Construction Code*; ESTIF: Brussel, Belgium, 2013.
9. HULC Unified Leader—Calener tool. Regulation FOM/1635/2013, September 10th (BOE September 12th) by which the Basic Document HE "Energy Saving" of the Spanish Building Construction Code is updated. Available online: https://www.codigotecnico.org/index.php/menu---recursos/menu---aplicaciones/282---herramientaunificada---lider---calener.html (accessed on 9 September 2017).
10. Passive House Database. Available online: https://passivehouse---database.org/index.php?lang=en (accessed on 7 May 2019).
11. Ridley, I.; Clarke, A.; Bere, J.; Altamirano, H.; Lewis, S.; Durdev, M.; Farr, A. The monitored performance of the first new London dwelling certified to the Passive House standard. *Energy Build.* **2013**, *63*, 67–78. [CrossRef]
12. Rekstad, J.; Meir, M.; Murtnes, E.; Dursun, A. A comparison of the energy consumption in two passive houses, one with a solar heating system and one with an air—Water heat pump. *Energy Build.* **2015**, *96*, 149–161. [CrossRef]
13. Valladares Rendón, L.G.; Schmid, G.; Lo, S.L. Review on energy savings by solar control techniques and optimal building orientation for the strategic placement of façade shading systems. *Energy Build.* **2017**, *140*, 458–479. [CrossRef]
14. Mahdavi, A.; Doppelbauer, E.M. A performance comparison of passive and low—Energy buildings. *Energy Build.* **2010**, *42*, 1314–1319. [CrossRef]
15. Bradke, H.; Weidemann, M.F.; Som, O.; Mannsbar, W.; Cremer, C.; Dreher, C.; Ruhland, S. *Improving the Efficiency of R&D and the Market Diffusion of Energy Technologies*; Springer Science & Business Media: Berlin, Germany, 2009.
16. Javier, N.G.; César, B.F. *Architectural and Constructive Environmental Conditioning Techniques*; Leria, M., Ed.; Munil la- Leria: Madrid, Spain, 1997.
17. Al Obaidi, K.M.; Ismail, M.; Rahman, A.M.A. Passive cooling techniques through reflective and radiative roofs in tropical houses in southeast asia: A literature review. *Front. Archit. Res.* **2014**, *3*, 283–297. [CrossRef]
18. Gagliano, A.; Detommaso, M.; Nocera, F.; Patania, F.; Aneli, S. The retrofit of existing buildings through the exploitation of the green roofs—A simulation study. *Energy Procedia* **2014**, *62*, 52–61. [CrossRef]

19.   Gagliano, A.; Nocera, F.; D'Amico, A.; Spataru, C. Geographical information system as support tool for sustainable Energy Action Plan. *Energy Procedia* **2015**, *83*, 310–319. [CrossRef]
20.   Pérez, G.; Coma, J.; Martorell, I.; Cabeza, L.F. Vertical greenery systems (VGS) for energy saving in buildings: A review. *Renew. Sustain. Energy Rev.* **2014**, *39*, 139–165. [CrossRef]
21.   Ubinas, R.E.; Montero, C.; Porteros, M.; Vega, S.; Navarro, I.; Castillo Cagigal, M.; Gutiérrez, A. Passive design strategies and performance of net energy plus houses. *Energy Build.* **2014**, *83*, 10–22. [CrossRef]
22.   Aksoy, U.T.; Inalli, M. Impacts of some building passive design parameters on heating demand for a cold region. *Build. Environ.* **2006**, *41*, 1742–1754. [CrossRef]
23.   Costanzo, V.; Donn, M. Thermal and visual comfort assessment of natural ventilated office buildings in Europe and North America. *Energy Build.* **2017**, *140*, 210–223. [CrossRef]
24.   Nocera, F.; Faro, A.L.; Costanzo, V.; Raciti, C. Daylight performance of classrooms in a mediterranean school heritage building. *Sustainability* **2018**, *10*, 3705. [CrossRef]
25.   Feist, W.; Schnieders, J.; Dorer, V.; Haas, A. Re—inventing air heating: Convenient and comfortable within the frame of the passive house concept. *Energy Build.* **2005**, *37*, 1186–1203. [CrossRef]
26.   Passive House requirements. Available online: https://passivehouse.com/02_informations/02_passive---house---requirements/02_passive---house---requirements.htm (accessed on 7 May 2019).
27.   Pitts, A. Passive house and low energy buildings: Barriers and opportunities for future development within UK practice. *Sustainability* **2017**, *9*, 272. [CrossRef]
28.   Passive House Institute (PHI). 2015. Available online: http://passiv.de/en (accessed on 8 November 2015).
29.   Fokaides, P.A.; Christoforou, E.; Ilic, M.; Papadopoulos, A. Performance of a Passive House under subtropical climatic conditions. *Energy Build.* **2016**, *133*, 14–31.
30.   Chen, X.; Yang, H.; Lu, L. A comprehensive review on passive design approaches in green building rating tools. *Renew. Sustain. Energy Rev.* **2015**, *50*, 1425–1436. [CrossRef]
31.   Clarke, J. A vision for building performance simulation: A position paper prepared on behalf of the IBPSA board. *J. Build. Perform. Simul.* **2015**, *8*, 39–43. [CrossRef]
32.   Anderson, K. *Design Energy Simulation for Architects: Guide to 3D Graphics*; Routledge: Abingdon, UK, 2014.
33.   Feist, W. *Passive House Planning Package 2007. Requeriments for Quality—Approved Passive Houses*; Passive House Institute: Darmstad, Germany, 2010.
34.   Gratia, E.; De Herde, A. Design of low energy office buildings. *Energy Build.* **2003**, *35*, 473–491. [CrossRef]
35.   Holm, D. Building thermal analyses—What the industry needs—The architects perspective. *Build. Environ.* **1993**, *28*, 405–407. [CrossRef]
36.   Soebarto, V.; Hopfe, C.; Crawley, D.; Rawal, R. *Capturing the Views of Architects about Building Performance Simulation to be Used During Design Processes*; Loughborough University: Loughborough, UK, 2015.
37.   De Souza, C.B. A critical and theoretical analysis of current proposals for integrating building thermal simulation tools into the building design process. *J. Build. Perform. Simul.* **2009**, *2*, 283–297. [CrossRef]
38.   Alsaadani, S.; De Souza, C.B. Of collaboration or condemnation? exploring the promise and pitfalls of architect—consultant collaborations for building performance simulation. *Energy Res. Soc. Sci.* **2016**, *19*, 21–36. [CrossRef]
39.   Papamichael, K.; Chauvet, H.; LaPorta, J.; Dandridge, R. Product modeling for computer—aided decision—making. *Autom. Constr.* **1999**, *8*, 339–350. [CrossRef]
40.   Papamichael, K. Decision making through use of interoperable simulation software. In Proceedings of the Building Simulation 97 5th International IBPSA Conference, Prague, Czech Republic, 8–10 September 1997.
41.   Papamichael, K. Application of information technologies in building design decisions. *Build. Res. Inf.* **1999**, *27*, 20–34. [CrossRef]
42.   Donn, M.R. Simulation of Imagined Realities: Environmental Design Decision Support Tools in Architecture. Ph.D. Thesis, Victoria University of Wellington, Wellington, New Zealand, 2004.
43.   Soebarto, V.I. Teaching an Energy Simulation Program in an Architecture School: Lessons Learned. Presented at the IBPSA 2005—International Building Performance Simulation Association, Montréal, QC, Canada, 15–18 August 2005; pp. 1147–1154.
44.   Bleil de Souza, C.; Tucker, S. Thermal simulation software outputs: A conceptual data model of information presentation for building design decision—Making. *J. Build. Perform. Simul.* **2016**, *9*, 227–254. [CrossRef]
45.   Hamby, D. A review of techniques for parameter sensitivity analysis of environmental models. *Environ. Monit. Assess.* **1994**, *32*, 135–154. [CrossRef]

46. Tomović, R. *Sensitivity Analysis of Dynamic Systems*; McGraw—Hill: New York, NY, USA, 1964.

47. Shneiderman, B. *The Eyes have it: A Task by Data Type Taxonomy for Information Visualizations. The Craft of Information Visualization*; Elsevier: Amsterdam, The Netherlands, 2003; pp. 364–371.

48. Cerd, I. Final report rehenergia project. Available online: http://www.jpi-oceans.eu/csa-oceans (accessed on 7 May 2019).

49. SNSI INE. Population and housing censuses. Tech. Rep. INE. 2001. Available online: http://www.ine.es/ (accessed on 7 May 2019).

50. DesignBuilder. Available online: www.desingbuilder.co.uk (accessed on 7 May 2019).

51. TABULA. 2016. Available online: http://episcope.eu/building---typology/webtool/ (accessed on 7 May 2019).

52. Gagliano, A.; Patania, F.; Nocera, F.; Galesi, A. Performance assessment of a solar assisted desiccant cooling system. *Therm. Sci.* **2014**, *18*, 563–576. [CrossRef]

53. Fichera, A.; Frasca, M.; Palermo, V.; Volpe, R. An optimization tool for the assessment of urban energy scenarios. *Energy* **2018**, *156*, 418–529. [CrossRef]

54. Evola, G.; Gagliano, A.; Fichera, A.; Marletta, L.; Martinico, F.; Nocera, F.; Pagano, A. UHI effects and strategies to improve outdoor thermal comfort in dense and old neighbourhoods. *Energy Procedia* **2017**, *134*, 692–701. [CrossRef]

55. Team, T.P. Synthesis report SR1: Use of building typologies for energy performance assessment of national building stocks, existent experiences in European countries and common approach. Tech. Rep. European Research Project. 2010. Available online: http://episcope.eu/fileadmin/episcope/public/docs/reports/EPISCOPESR1NewBuildingsInTypologies.pdf (accessed on 7 May 2019).

56. Government of Spain. Ministry of Industry, Tourism and Commerce. Institute for diversification and energy saving. Available online: https://www.idae.es/uploads/documentos/documentos_CALENER_07_Escala_Calif_Energetica_A2009_A_5c0316ea.pdf (accessed on 7 May 2019).

57. Depecker, P.; Menezo, C.; Virgone, J.; Lepers, S. Design of buildings shape and energetic consumption. *Build. Environ.* **2001**, *36*, 627–635. [CrossRef]

58. Littlefair, P. Passive solar urban design: Ensuring the penetration of solar energy into the city. *Renew. Sustain. Energy Rev.* **1998**, *2*, 303–326. [CrossRef]

 *sustainability*

Article

The Evaluation of Single-Family Detached Housing Units in Terms of Integrated Photovoltaic Shading Devices: The Case of Northern Cyprus

John Emmanuel Ogbeba * and **Ercan Hoskara**

Department of Architecture, Faculty of Architecture, Eastern Mediterranean University, North Cyprus, 99450 Famagusta, Mersin 10, Turkey; ercan.hoskara@emu.edu.tr
* Correspondence: john.ogbeba@emu.edu.tr

Received: 6 December 2018; Accepted: 15 January 2019; Published: 23 January 2019

Abstract: In this paper, we evaluate passive and active strategies that can be used in solving the heating problems in the residential sector of Northern Cyprus. In doing so, we propose the use of photovoltaics as a shading device (PVSD). PVSD is known to produce clean energy from solar radiation and it also reduces the energy consumed for cooling. We use an empirical method to evaluate the performance of a typical family detached dwelling in Famagusta, Cyprus. The simulation result derived from the study indicates that the strategic use of PVSDs for openings oriented towards the east, west, and south can reduce its energy consumption by almost 50% in three months of the year and cut down up to 400 kWh of energy consumption through the year, thus raising the comfort level of the building by about 20%. It will also generate nearly 2800 W that can provide up to 50% of the electricity demand.

Keywords: renewable energy integration; shading devices; BIPV; thermal comfort; energy consumption; Northern Cyprus

---

1. Introduction

Currently, the world is using fossil fuels at an alarming rate that not only will strain the sources in the near future but will result in a great amount of pollution as well. In fact, the effect of the over dependency on the use of fossil fuel for energy generation is becoming more evident in our world today. Global warming, climate change, large carbon footprints, decreasing fossil energy sources, etc., are some of the glaring effects resulting from the over use of fossil fuels.

According to the Statistical Review of World Energy (June 2016), about 33% of the total energy consumed globally in the year 2015 was from oil, as shown in Figure 1 [1]. Buildings have been identified as one of the largest energy consumers in the world today, accounting for approximately 40% of Europe's energy consumption. This makes buildings a major contributor to greenhouse emissions and climate change. In the case of Northern Cyprus, 70% of the energy generated is consumed by buildings (KIB-TEK) [2,3]. Northern Cyprus has no oil or gas reserves, and imports the gasoline and oil used in its entire energy generation. The state-run utility company known as the Turkish Electricity Authority (KIB-TEK) and the private sector's AKSA energy are responsible for the generation, sales, and distribution of electricity; producing a total of about 403.2 MW (KIB-TEK), as shown in Table 1 [2]. Industrial consumption of electricity is quite low in Northern Cyprus, therefore, the bulk of the electricity generated is domestically consumed; used mainly for heating and cooling buildings, or for powering electronic devices and lighting. In Cyprus, a number of residential buildings are constructed without paying attention to certain basic bio-climatic principles which eventually lead to an over-dependence on oversized active systems for cooling and heating of spaces [3]. In addition, the lack of insulation in most of the buildings' walls relatively increases the thermal discomfort within the interior spaces.

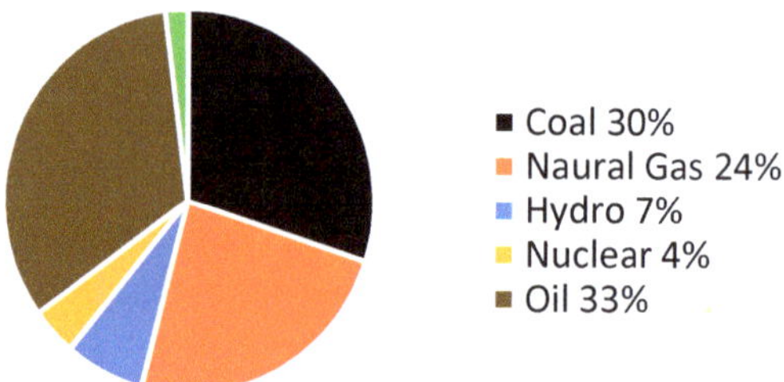

Figure 1. World total primary energy consumption by fuel in 2015 [1].

Table 1. Total installed capacity in Northern Cyprus at the first half of 2018 [2].

Location	Type of Technology	Number and Sizeof Units	Fuel	Total Installed Capacity
Teknecik	Steam Turbine	2 × 60 MW	Heavy fuel oil	120 MW
Teknecik	Gas Turbine	20 MW + 10 MW	Diesel	30 MW
Teknecik	Diesel Generator	8 × 17.5 MW	Diesel	140 MW
Dikmen	Gas Turbine	20 MW	Diesel	20 MW
Kalecik	Diesel Generator	8 × 17 MW	Diesel	136 MW
Kalecik	Steam Turbine	6 MW	Heavy fuel oil	6 MW
Serhat	Solar	1.2 MW	-	1.2 MW
Total				453.2 MW
Total without Gas Turbines				403.2 MW

On the other hand, there is an increasing awareness of the environmental problems caused by the use of non-renewables [4], thus leading to a growing interest in the use of cleaner and renewable sources of energy such as wind energy, solar energy, etc. Besides, it has become obvious that the use of energy from renewable sources is the most effective way of combating climate change. Solar energy, for instance, has been identified as the most popular source of renewable energy today. The energy from solar irradiation on the earth can meet the daily energy consumption needs of mankind several thousand times over [5]. There are two basic categories of solar energy: passive and active. Passive solar energy is put into practice as a design strategy to accomplish direct or indirect space heating, daylighting, etc. [6]. Active solar energy is implemented through technical installations, such as solar collectors and photovoltaic (PV) panels.

Today, one of the most widespread technologies of renewable energy generation is the use of photovoltaic (PV) systems which convert sunlight into electrical energy. Altin opined that "solar energy is the most suitable and easiest renewable energy source that can be used in buildings" and the most popular renewable energy source [7,8].

Several studies have been carried out globally that have focused on finding practical ways and methods for conserving and optimizing energy within buildings. Some of the major practical strategies used in conserving energy within building spaces include; good cross ventilation, allowing daylight into the building, and cutting down on the excessive use of electricity [9–11]. Within the Mediterranean region, several studies have also been conducted on how to use both passive and active measures to improve the indoor environment [12,13]. Moreover, lots of researchers have endeavored to emphasize the significance of housing quality and the need to establish well-designed buildings that are energy efficient [12–17]. Therefore, since no two residential buildings are exactly the same, it is quintessential to analyze the energy saving potentials of houses and their construction techniques at an individual household scale and the result can then be replicated on a wider regional scale [18].

In this paper, we evaluate passive and active strategies that can be used in solving the heating problems in the residential sector of Northern Cyprus. In doing so, the best design strategies for solving the overheating problem of a single-family detached house were identified and presented. The study also presented the best way to integrate PVs as shading devices. This article is divided into six sections. The first section provides a brief introduction and the aim of the study. In the second section, building-integrated photovoltaics (BIPV) and photovoltaics as a shading device (PVSD) are discussed. The third section of this study presents the context of the study location, while the methodology for the study is discussed in the fourth section. Sections 5 and 6 contain the results/findings and conclusion of the research, respectively.

2. BIPV and PVSD

BIPV: The term building-integrated photovoltaics (BIPVs) refers to the integration of photovoltaic panels into the building skin [19], with the dual roles of replacing building components and of simultaneously serving as electricity generators [20–22]. There have also been possibilities of BIPV products being used as façade, windows, shading elements or awnings, curtain walls, and roofs [23–27].

PVSDs: Architects always endeavor to improve the quality of interior spaces by creating openings for daylighting and views to the outside. This most often can result in excessive solar gains. Shading devices are used as a passive strategy to reduce this effect. As indicated by Schittich, shading systems help: prevent overheating in indoor spaces, adapt to different climates, reduce energy consumption for cooling and heating, provide a glare-free environment, and help to control the ventilation of indoor spaces [28]. Pester and Crick mention four different ways photovoltaic panels can be integrated on the building skin [29], as shown in Figure 2.

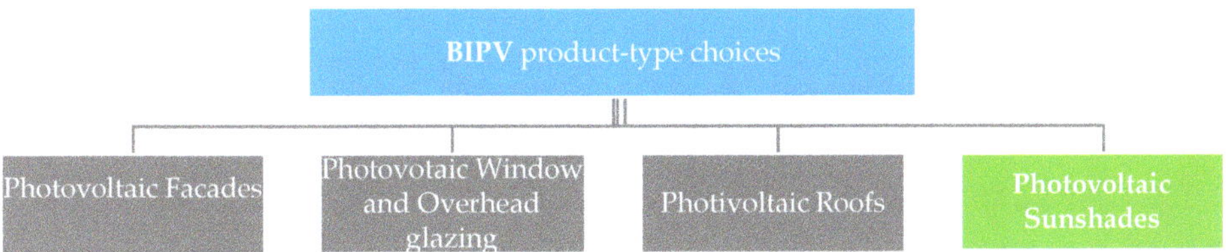

Figure 2. Four types of building-integrated photovoltaic (BIPV) product choices [29].

Although most literature focuses on the first three, the photovoltaic shading device is becoming more and more relevant for integration in buildings. The use of PVSDs is in fact more technical than the other three [30]. The use of PVSDs on buildings have economic and environmental advantages as shown in Table 2.

Table 2. Advantages of photovoltaics as shading devices (PVSDs).

Sustainability Dimension	Benefits	References
Environmental aspects	Ability to produce clean energy by converting unwanted solar radiation into electrical energy. There is no side effect to the ecosystem. The cooling load is reduced when PVSDs are introduced. When PVs are used as shading device, they generate energy and serve as protection from glare, hence improving the visual comfort of the users of the space, enhancing specific architectural expression through the application of colored and semi-transparent PV products.	[31–34].
Economical aspects	The energy is generated in situ and not centrally, there are savings in terms of the cost of distribution and greenfield. The cost of material is saved since the PV is serving as both an energy generator and as a building component.	[34,35]

The outcomes of the shading study of Aksugur, in his paper on the potential of passive cooling strategies in Cyprus, show that rooms with windows which are located on the south side of a building become excessively hot in summer [36]. Oktay's evaluation of vernacular housing in Northern Cyprus suggests that orienting living spaces towards the south, as well as designing a narrow building plan with windows on the opposite sides for cross-ventilation, are essential passive design strategies that can be used in Northern Cyprus [37]. The primary objective of using passive and active design strategies is to create a soothing micro-climate which is cool in the summers and warm in the winters [38–42].

Over the years, several strategies have been proposed that promote the use of PV systems within the residential housing stock. According to James et al., "declining module cost, growing consumer interest in solar energy and policy schemes" are some of the key factors to put into consideration [43]. Other major factors that influence the generation and consumption of energy include the location and climate data of the building, the orientation of the building, installed materials, building design, and the selection of the technical systems [44–46].

3. The Context of the Study

The single-family housing type case study was located in Famagusta, Cyprus, as shown in Figure 3. Cyprus is the third largest island in the Mediterranean Sea. The city of Famagusta is a coastal town and is on the east side of the island, with a seven-meter elevation above sea level [47]. The population of Famagusta is 42,526 according to the 2006 census [48].

The climate in Famagusta is a transitional one; it lies between a composite and a hot, humid climate, however, because of its close proximity to the Mediterranean Sea, it has a hot and humid climate [49,50]. According to the Cyprus meteorological station report about Famagusta, the temperature of Famagusta rises to more than 30 °C in the summer months and drops to a low of 3 °C in the winter months, as shown in Figure 4. The relative humidity for the city of Famagusta is between 33% to 72% in the different months of the year. January records the highest and October the lowest humidity in the year. Famagusta has an average of 9 h of sunlight each day, as shown in Figure 4, and an average of 3328 h of sunlight in a year. The city also experiences an average of 403.5 mm of rainfall each year and an average of 33.6 mm each month [49].

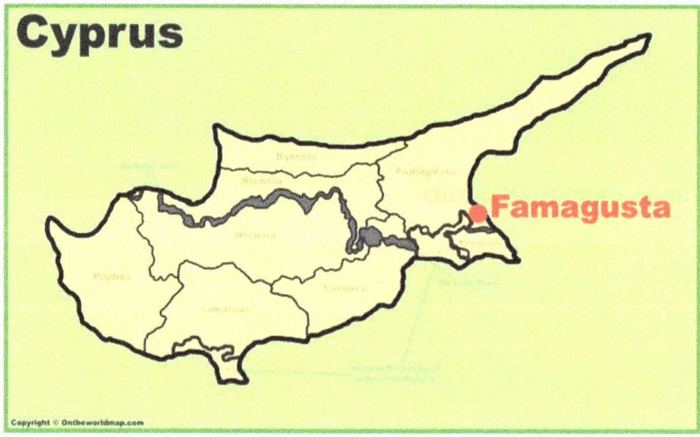

Figure 3. Location of Famagusta city in Cyprus [51].

The most comfortable months of the year are the months of April, May, October, and November, while the months of December, January, February, and March require heating. The summer months of June, July, August, and September require cooling and ventilation [52]. According to [53], Famagusta has high solar energy during the winter (5.26 kWh/m^2/day), which rises to 7.12 kWh/m^2/day during the summer season, as shown in Figure 6. In addition, the total solar radiation in Famagusta usually rises from 6 MJ/m^2 in December to a high of about 24 MJ/m^2 in June and July. Besides, the energy generation rises from 70 W/m^2 in December to 280 W/m^2 in June and July [54,55].

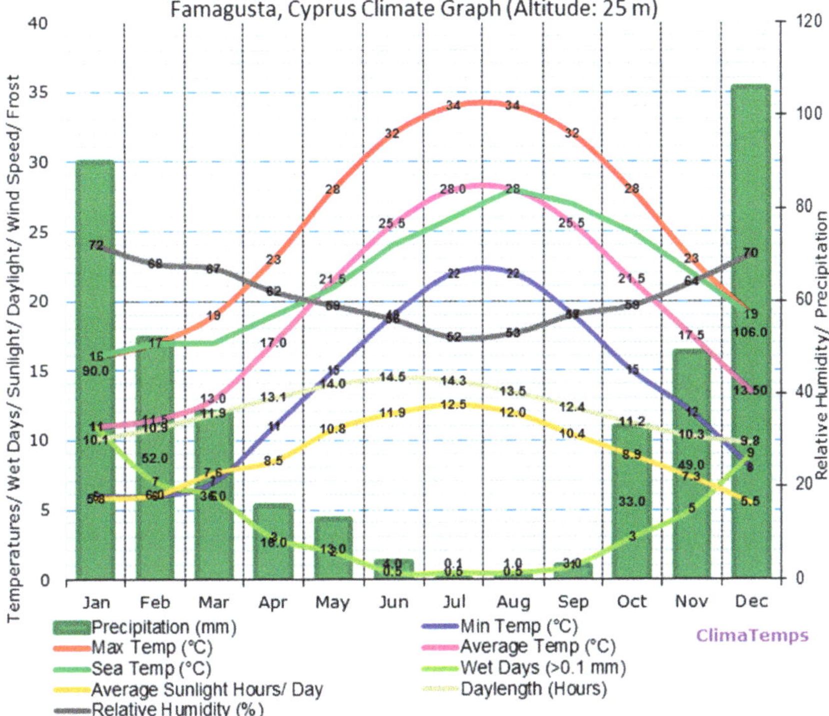

Figure 4. Annual climate graph of Famagusta [53].

Figures 5 and 6 illustrate that the countrywide solar potential of Cyprus is the highest in Europe. Moreover, studies have shown that a polycrystalline or monocrystalline solar PV system with nominal power of 1 kW installed in the coastal region of Cyprus, with a 27° still angle of the panels and south direction, produces on average more than 1500 kWh per year throughout the first 20 years of its operation [56]. This is a clear indication of why the government should invest in solar energy as an alternative source of energy.

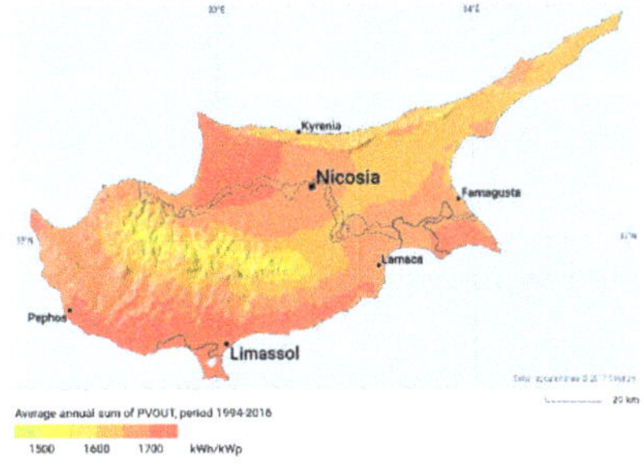

Figure 5. Photovoltaic power potential in Northern Cyprus [57].

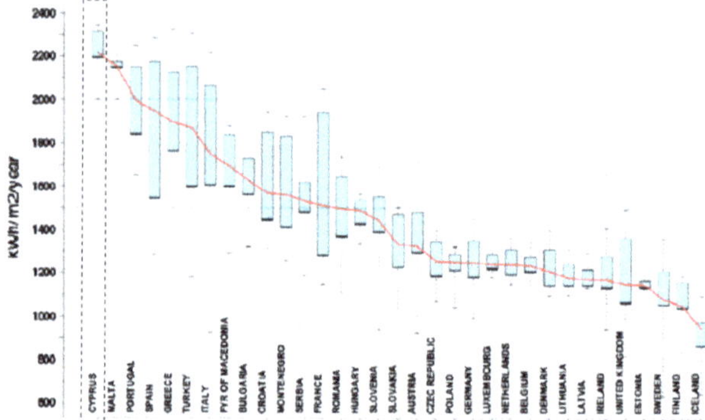

Figure 6. Photovoltaic power potential in Northern Cyprus [56].

4. Materials and Methodology

To achieve the objectives of this research, both qualitative and quantitative methodology were employed, as shown in Figure 7. The qualitative method included a wide-ranging survey of relevant literature, while the quantitative method involved a model set up. Firstly, a typical single-family house was chosen for the simulation. A building simulation model was generated to reproduce the interior and outside of the house, as well as the materials used for the construction of the building. The computer program DesignBuilder was used. This software uses EnergyPlus™, a simulation software developed by the US Department of Energy (DOE) to simulate heat transfer processes, climate conditions, and other factors relating to energy consumption in buildings. EnergyPlus™ is a whole building energy simulation program that researchers and professionals in the building industry use to model both energy consumption for heating, cooling, ventilation, lighting, and plug and process loads—and water use in buildings [58,59].

Furthermore, the Climate Consultant program and the weather file of Famagusta (epw) was used to develop the best strategies for both passive and active design that would create a unique list of design guidelines for the building. The building was simulated with and without shading devices, and finally with the shading devices replaced with PVs. The ASHRAE Standard 55 was also used as the comfort model in the Climate Consultant simulation. Milne describes Climate Consultant as a graphic-based computer program that helps users create more energy efficient buildings, each of which is uniquely suited to its particular location in the world [60].

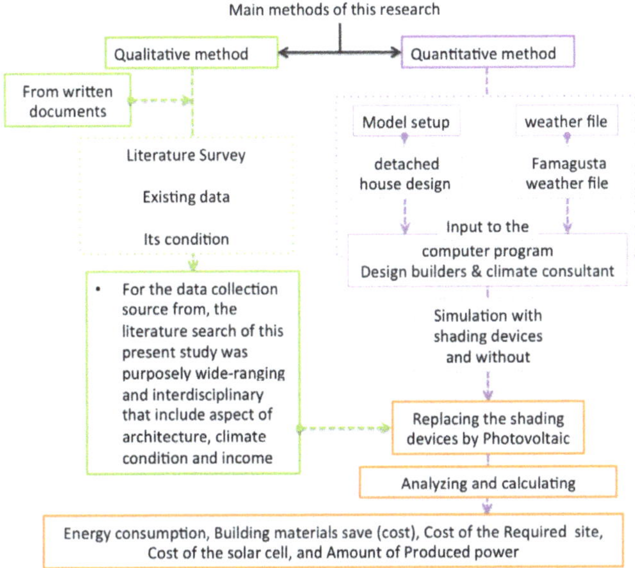

Figure 7. Illustration diagram of process and methodology of the research (drawn by authors).

Building and PVSD Case Study

For the model setup, a common single-family detached dwelling with a living room, a garage, a kitchen, two bathrooms, and five bedrooms was evaluated using DesignBuilder software. The house has a south orientation, a flat roof, and a total floor area of 149 m^2. The details about the building model are shown in Figure 8 and Table 3.

(a)

(b)

Figure 8. (**a**) Ground first floor of the building model; (**b**) elevation of the building.

Table 3. Information about the single-family building model.

s/n	Building Information	Description
1	Project	Residential building
2	Type	Single family detached house
3	Area of building	149 m^2
4	Climatic region	Mediterranean climate
5	Orientation:	South
6	Number of floors	Two floors
7	Window material	PVC
8	Window configuration	Double glazing
9	Windows type	3 mm clear glass + 6 mm air gap + 3 mm clear glass
10	Level above ground	1 m
11	Orientation of the openings	South, east, west, and north
12	Outside shading device	Louver and cantilever
13	Door material	Wood
14	Wall type	Bearing and partition
15	HVAC system	Slit unit

EnergyPlus operational and occupancy schedules default values for residential buildings were used for the simulation. The building's energy demand includes; lighting, domestic hot water (DHW), air-conditioning system, and other household appliances. The DesignBuilder software's default residential operational schedule was used to determine the building's electricity requirements. The input parameters and boundary conditions for the EnergyPlus simulation used to determine the light, DHW, and other appliances are given in Table 4. The heat transfer coefficients of the construction elements of the building were obtained from [61] for the heating and cooling load calculations and are tabulated in Table 5.

Table 4. Parameters for residential electricity demand. DHW: domestic hot water.

Lighting	Illuminance (lux)	150
DHW	Consumption rate (l/m2-day)	0.72
	Delivery temperature (C)	65
Equipment	Unit consumption (W/m2)	3

Table 5. Heat transfer coefficient (U) for construction components [61].

City	U–Value (W/m^{2}k)			
	Wall	Roof Ceiling	Floor	Windows
Famagusta Northern Cyprus	0.56	0.67	0.44	0.8

5. Result and Findings

This section of the paper presents the results from the simulation of the case study.

5.1. Sun Shading Chart

From the Climate Consultant program, the number of hours that shading is required for the simulated building was derived. The result of the simulation presented in Figures 9 and 10 show the annual number of hot, cold, and comfortable hours in the simulated building. Between the winter and spring months (21 December to 21 June), shade would be needed for 174 h, the sun will be needed for 1590 h, and 748 h are the comfortable hours within this period of the year. While for the summer months as shown in Figure 10 (21 June to 21 December), shading will be required for 1203 h, direct solar radiation for heating is required in the building for 499 h, and the remaining 896 h are the comfortable hours of the year.

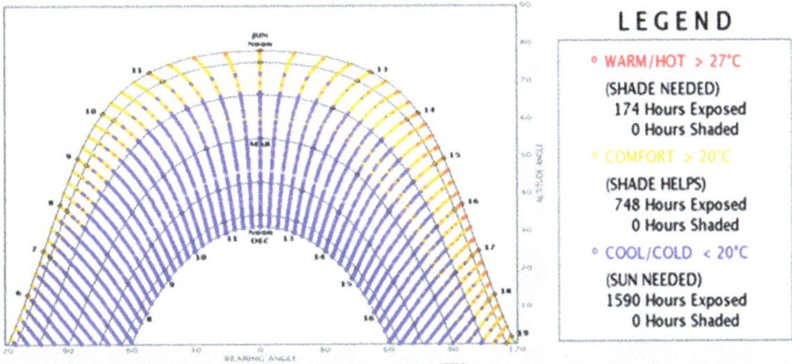

Figure 9. Sun shading chart in winter and spring (21 December to 21 June).

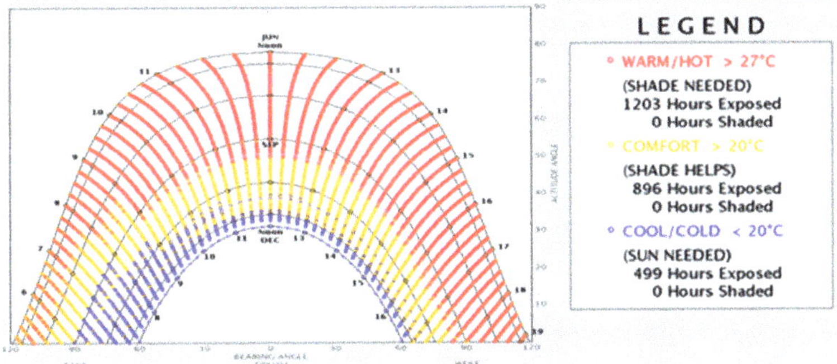

Figure 10. Sun shading chart in summer and fall (21 June to 21 December).

Furthermore, from Figures 9 and 10, we can deduce that the two-floor single-family house, which is classified as a low-rise residential building, needs shading devices for a total of 1218 h in a year.

5.2. Psychrometric Chart Results

One very important output of the Climate Consultant software is the psychrometric chart. Beyond just representing the climatic data, the psychometric chart helps to organize the information in a way that is plain and easy for people to understand the influence of climate on the immediate environment. The Comfort Zone, as shown in Figure 11, is the psychrometric chart for the simulated house. It can be clearly seen that the area of the comfort zone for the building is quite small (16.8%), implying that a large amount of energy would be needed for heating and cooling. Therefore, very good passive and active design strategies need to be developed to solve the heating problem.

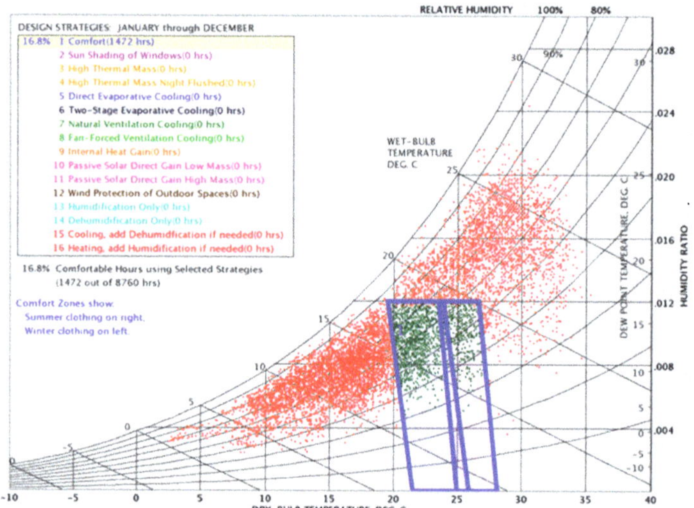

Figure 11. Psychrometric chart for Famagusta: Comfort Zone.

Figure 12 presents the best design strategies for building envelopes in Famagusta. The strategies are able to modify or filter extreme external climate conditions to create comfortable indoor environments in Famagusta.

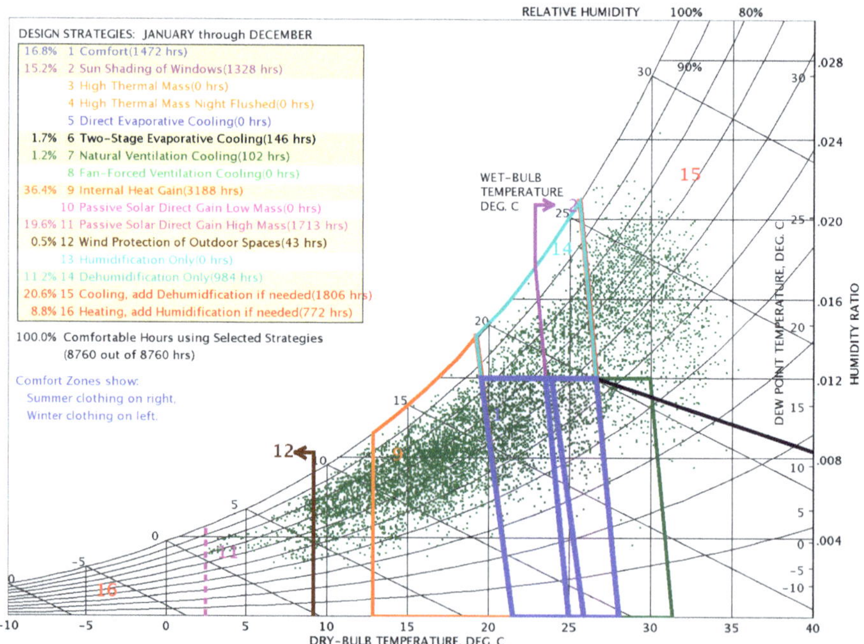

Figure 12. Psychrometric chart for Famagusta.

The design strategies for the simulated building are explained as follows:

- Sun shading of windows: as presented in Figure 15, shading on the chart takes up to 15.2% that is about 1328 h of the year. Through the use of shading devices, 1328 uncomfortable hours are converted to comfortable hours.
- Two stage evaporative cooling: in the two-stage evaporative cooling strategy, first a thermal converter is used to reduce the temperature, then the comfort condition is applied by direct evaporative cooling. This process makes up 1.7% (146 h annually).
- Natural ventilation cooling: natural ventilation is required for cooling for about 102 h of the year (1.2%).
- Internal heat gain: 36.4% of thermal comfort can be achieved by internal heat gained from within the building from artificial lighting, electrical equipment, and indoor activities by occupants. This is about 3188 h of the year.
- Passive solar direct gain high mass: the is the number of hours in the year where thermal comfort is achieved through passive solar gain. This includes a total of about 1713 h (19.6%).
- Wind protection of outdoor spaces: in this segment of the chart, building wind protection by outdoor elements such as plants is required to achieve the comfort conditions. This includes 0.5% making up a total of 43 h of the year.
- Dehumidification only: dehumidification is required to achieve thermal comfort in the building for a total of about 984 h of the year, making up 11.2%.
- Cooling, add humidification if needed: to achieve comfort, this strategy requires both cooling and humidification at the same time. This includes a total of 1806 h of the year (20.6%).
- Heating, add humidification if needed: to achieve comfort, this strategy requires both humidification and increasing air temperature by mechanical heating. This includes a total of 772 h of the year (8.8%).

5.3. Solar Shading Performance Result

Solar shading performance was also assessed using Climate Consultant. Shading elements work differently based on their orientation. Shading device strategies are usually tailored towards the orientation of each window. Whilst some orientations are easy to shade, others are much more difficult as the sun can be almost direct-on at certain times of the day. The number of hours exposed to the sun that need to be shielded is also different based on the direction the façade is facing (north, south, east, and west). From the simulation, the types of shading elements that should be used are also different, as illustrated in Figures 13 and 14. The results from the simulation are that:

- Windows directly facing the south would need more shading from the sun, and the horizontal overhangs work better for southern façades. The east and west would require both vertical fins and horizontal overhanging used in the passive design strategies, while on the northern façade, shading is completely avoided as exposure to the sun is needed for the interior space of the building.

By having shading devices with PV integrated retrofits, shading can be provided, and electricity generated simultaneously, which is what this research seeks to achieve.

Having identified shading as one of the best strategies, the next step would be to identify what shading system best fits the orientation of the building. Through the use of Climate Consultant, this study has been able to identify the best shading strategy that best fits the single-family building location, as shown in Figure 13.

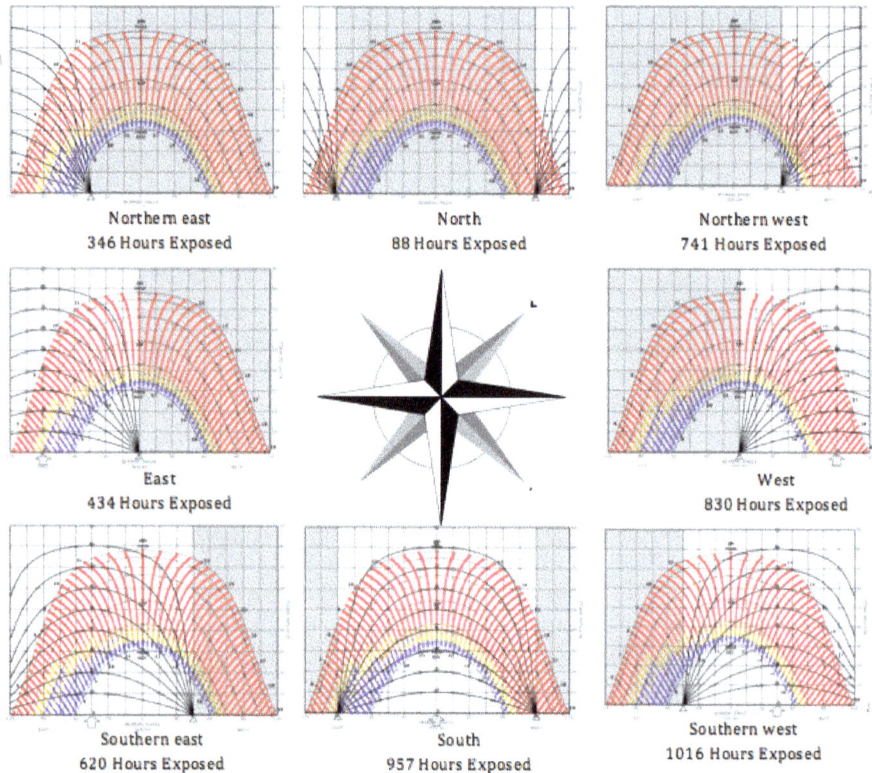

Figure 13. The Shading Calculator Overlay in eight different orientation cases by focusing on sun exposure during June 21 to December 21.

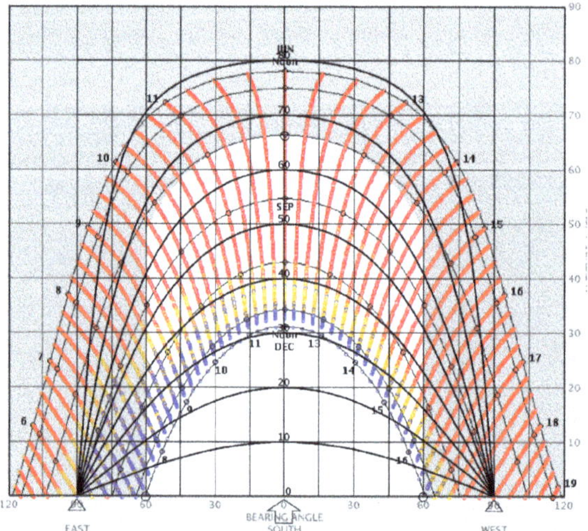

Figure 14. Overhang angle and vertical fin angle.

5.4. Cooling Design Strategy

The cooling design is aimed at solving the overheating problem of the building, especially in the summer months. Figures 15 and 16 show the cooling design calculation, the size of the cooling plan required in design during the summer condition, and the impact of the changing aspects of shading to help find a design solution.

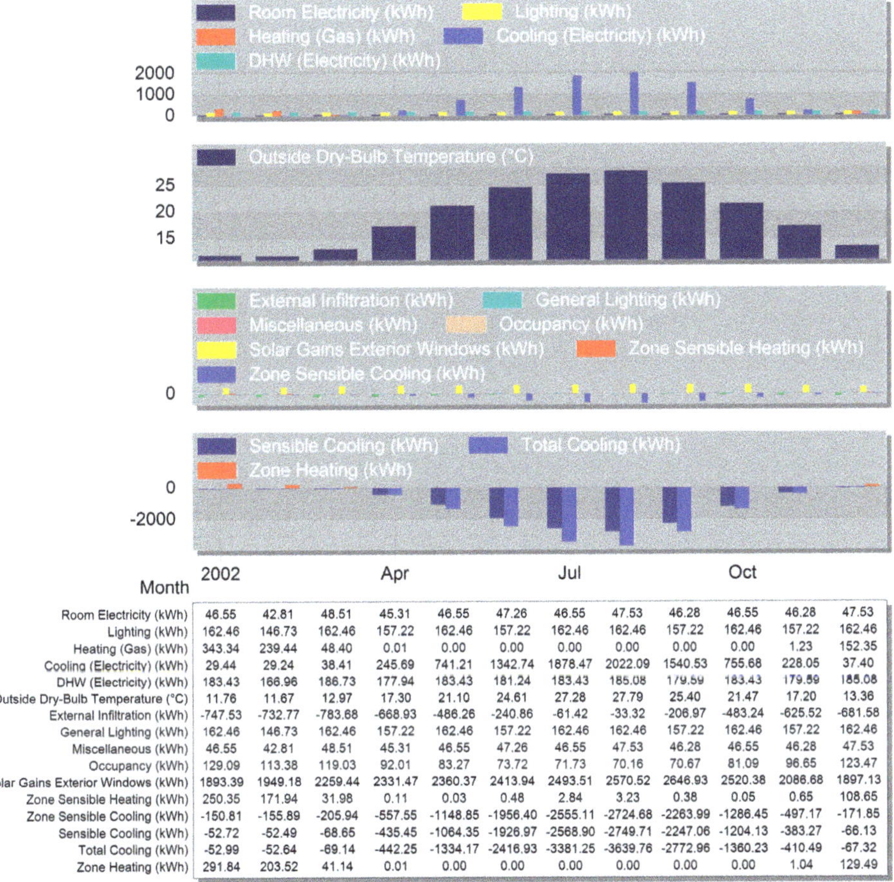

Month	2002			Apr			Jul			Oct		
Room Electricity (kWh)	46.55	42.81	48.51	45.31	46.55	47.26	46.55	47.53	46.28	46.55	46.28	47.53
Lighting (kWh)	162.46	146.73	162.46	157.22	162.46	157.22	162.46	162.46	157.22	162.46	157.22	162.46
Heating (Gas) (kWh)	343.34	239.44	48.40	0.01	0.00	0.00	0.00	0.00	0.00	0.00	1.23	152.35
Cooling (Electricity) (kWh)	29.44	29.24	38.41	245.69	741.21	1342.74	1878.47	2022.09	1540.53	755.68	228.05	37.40
DHW (Electricity) (kWh)	183.43	166.96	186.73	177.94	183.43	181.24	183.43	185.08	179.59	183.43	179.59	185.08
Outside Dry-Bulb Temperature (°C)	11.76	11.67	12.97	17.30	21.10	24.61	27.28	27.79	25.40	21.47	17.20	13.36
External Infiltration (kWh)	-747.53	-732.77	-783.68	-668.93	-486.26	-240.86	-61.42	-33.32	-206.97	-483.24	-625.52	-681.58
General Lighting (kWh)	162.46	146.73	162.46	157.22	162.46	157.22	162.46	162.46	157.22	162.46	157.22	162.46
Miscellaneous (kWh)	46.55	42.81	48.51	45.31	46.55	47.26	46.55	47.53	46.28	46.55	46.28	47.53
Occupancy (kWh)	129.09	113.38	119.03	92.01	83.27	73.72	71.73	70.16	70.67	81.09	96.65	123.47
Solar Gains Exterior Windows (kWh)	1893.39	1949.18	2259.44	2331.47	2360.37	2413.94	2493.51	2570.52	2646.93	2520.38	2086.68	1897.13
Zone Sensible Heating (kWh)	250.35	171.94	31.98	0.11	0.03	0.48	2.84	3.23	0.38	0.05	0.65	108.65
Zone Sensible Cooling (kWh)	-150.81	-155.89	-205.94	-557.55	-1148.85	-1956.40	-2555.11	-2724.68	-2263.99	-1286.45	-497.17	-171.85
Sensible Cooling (kWh)	-52.72	-52.49	-68.65	-435.45	-1064.35	-1926.97	-2568.90	-2749.71	-2247.06	-1204.13	-383.27	-66.13
Total Cooling (kWh)	-52.99	-52.64	-69.14	-442.25	-1334.17	-2416.93	-3381.25	-3639.76	-2772.96	-1360.23	-410.49	-67.32
Zone Heating (kWh)	291.84	203.52	41.14	0.01	0.00	0.00	0.00	0.00	0.00	0.00	1.04	129.49

Figure 15. Simulation results without shading on a monthly basis: Temperature, Heat Gains, and Energy Consumption—1, Building 1. EnergyPlus 1 Output 31 December (zone conditions reported for occupied periods, defined by schedule) Monthly Evaluation.

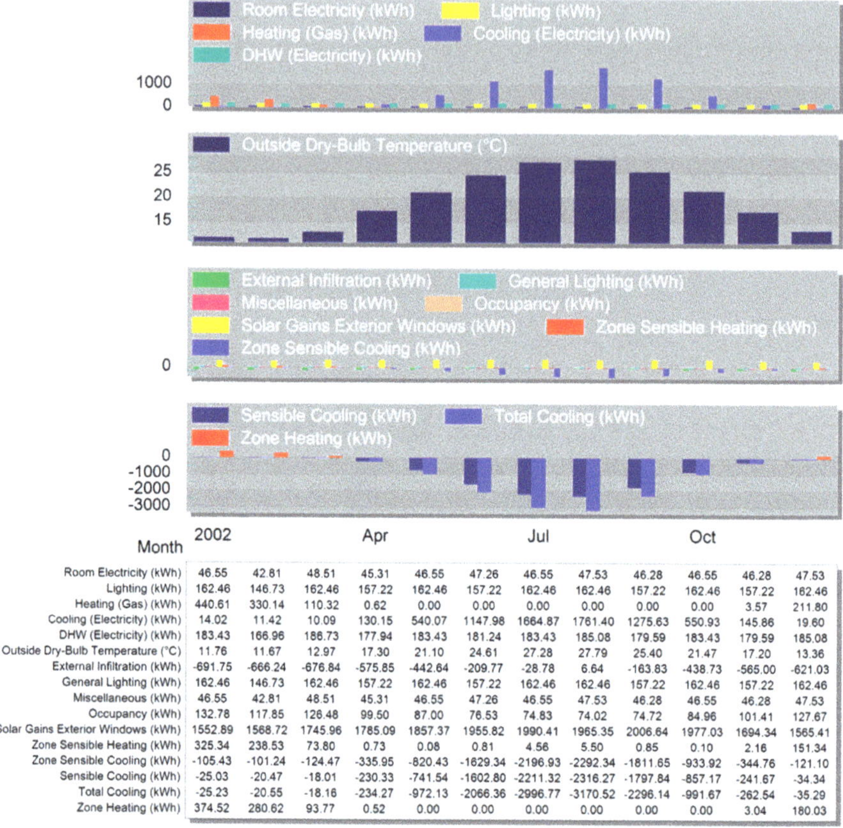

Month	2002			Apr			Jul			Oct		
Room Electricity (kWh)	46.55	42.81	48.51	45.31	46.55	47.26	46.55	47.53	46.28	46.55	46.28	47.53
Lighting (kWh)	162.46	146.73	162.46	157.22	162.46	157.22	162.46	162.46	157.22	162.46	157.22	162.46
Heating (Gas) (kWh)	440.61	330.14	110.32	0.62	0.00	0.00	0.00	0.00	0.00	0.00	3.57	211.80
Cooling (Electricity) (kWh)	14.02	11.42	10.09	130.15	540.07	1147.98	1664.87	1761.40	1275.63	550.93	145.86	19.60
DHW (Electricity) (kWh)	183.43	166.96	186.73	177.94	183.43	181.24	183.43	185.08	179.59	183.43	179.59	185.08
Outside Dry-Bulb Temperature (°C)	11.76	11.67	12.97	17.30	21.10	24.61	27.28	27.79	25.40	21.47	17.20	13.36
External Infiltration (kWh)	-691.75	-666.24	-676.84	-575.85	-442.64	-209.77	-28.78	6.64	-163.83	-438.73	-565.00	-621.03
General Lighting (kWh)	162.46	146.73	162.46	157.22	162.46	157.22	162.46	162.46	157.22	162.46	157.22	162.46
Miscellaneous (kWh)	46.55	42.81	48.51	45.31	46.55	47.26	46.55	47.53	46.28	46.55	46.28	47.53
Occupancy (kWh)	132.78	117.85	126.48	99.50	87.00	76.53	74.83	74.02	74.72	84.96	101.41	127.67
Solar Gains Exterior Windows (kWh)	1552.89	1568.72	1745.96	1785.09	1857.37	1955.82	1990.41	1965.35	2006.64	1977.03	1694.34	1565.41
Zone Sensible Heating (kWh)	325.34	238.53	73.80	0.73	0.08	0.81	4.56	5.50	0.85	0.10	2.16	151.34
Zone Sensible Cooling (kWh)	-105.43	-101.24	-124.47	-335.95	-820.43	-1629.34	-2196.93	-2292.34	-1811.65	-933.92	-344.76	-121.10
Sensible Cooling (kWh)	-25.03	-20.47	-18.01	-230.33	-741.54	-1602.80	-2211.32	-2316.27	-1797.84	-857.17	-241.67	-34.34
Total Cooling (kWh)	-25.23	-20.55	-18.16	-234.27	-972.13	-2066.36	-2996.77	-3170.52	-2296.14	-991.67	-262.54	-35.29
Zone Heating (kWh)	374.52	280.62	93.77	0.52	0.00	0.00	0.00	0.00	0.00	0.00	3.04	180.03

Figure 16. Simulation results with shading devices on a monthly basis: Temperature, Heat Gains, and Energy Consumption—1, Building 1. EnergyPlus 1 Output 31 December (zone conditions reported for occupied periods, defined by schedule) Monthly Evaluation.

Figure 18 shows that the amount of energy needed for cooling the sample building is very high, especially in the summer months of June, July, August, and September. Implementing the shading device strategies derived from the solar shading performance result of this research would lead to a more than 400 kWh reduction, especially in some of the summer months, and the entire annual cooling load reduced by almost half (total cooling load—Figure 16). From the simulation, the total energy consumption that will be reduced in the summer months of June, July, August, and September is 716.02 kWh. On the other hand, the heating load increases in winter but not in a significant measure. The increase in the heating load for winter months of December, January, February, and March amounts to 115.70 kWh. In this tradeoff, the reduction in energy consumption in the summer months still outweighs the increase in the winter months.

5.5. Total Energy Generated by the PVSDs and Cell Type

The type of solar cell used for the simulation is the monocrystalline cell. One meter squared of monocrystalline cell produces 160 watts as earlier mentioned. Since there is 20.4 m² available surface for integration, the PV shading device will generate a total of about 3264 watts of electricity and from due south with every 5°, the average deficiency drop is about 1.1% with the number of windows oriented to both east and west.

Total of surface area of shading devices on the ground floor that are exposed to the sun:

$$(1.8 + 1.4) \times 0.4 + (2.3 + 1.4) \times 0.4 + 3.2 \times 0.4 + 4.9 \times 0.4 + (4.4 + 2.0) \times 0.4 + (2.3 + 1.4) \times 0.4 = 10.08 \text{ m}^2 \quad (1)$$

Surface area of shading devices on the first floor that are exposed to the sun:

$$(2.6 + 1.9) \times 0.4 + (1.8 + 1.4) \times 0.4 + (2.3 + 1.4) \times 3.0 \times 0.4 + 2.0 \times 0.4 + 2.4 \times 0.4 +$$
$$(2.9 + 1.2) \times 0.4 + (2.8) \times 0.4 = 10.28 \text{ m}^2 \quad (2)$$

Total surface area = $10.08 + 10.28 = 20.36 \approx 20.4 \text{ m}^2$

Figures 17 and 18 show the layout and the perspective view of the simulated building. Figure 17a shows the building without the PVSD, while 17b shows the building with the PVSD. Figure 19 shows the PVSD system integrated into the simulated building.

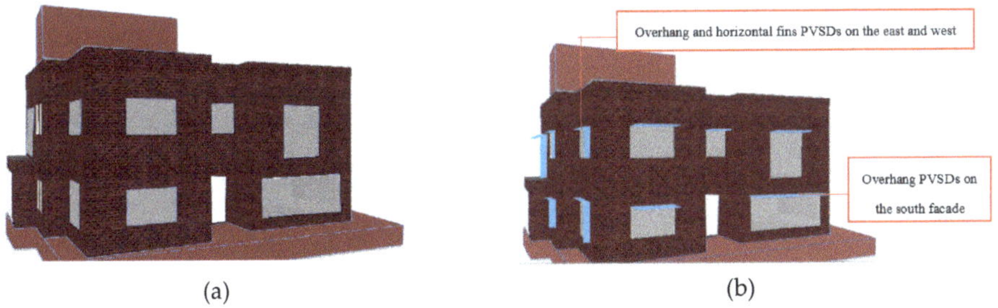

(a)　　　　　　　　(b)

Figure 17. (**a**) Without shading devices; (**b**) with shading.

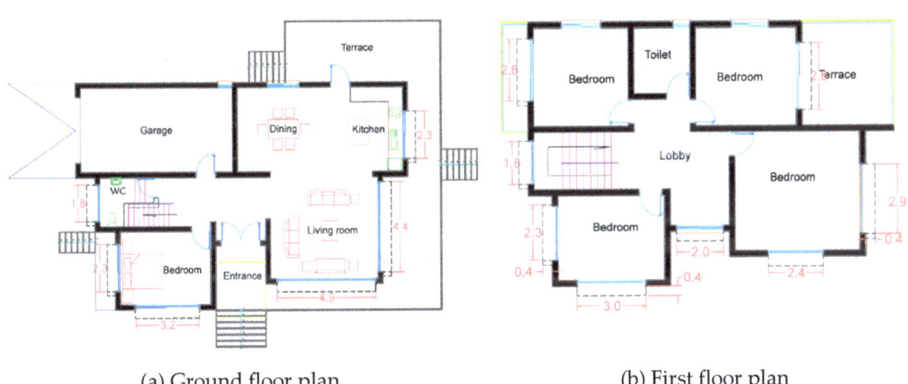

(a) Ground floor plan　　　　　　(b) First floor plan

Figure 18. Building layout with the shading device. (**a**) Ground floor plan, (**b**) First floor plan.

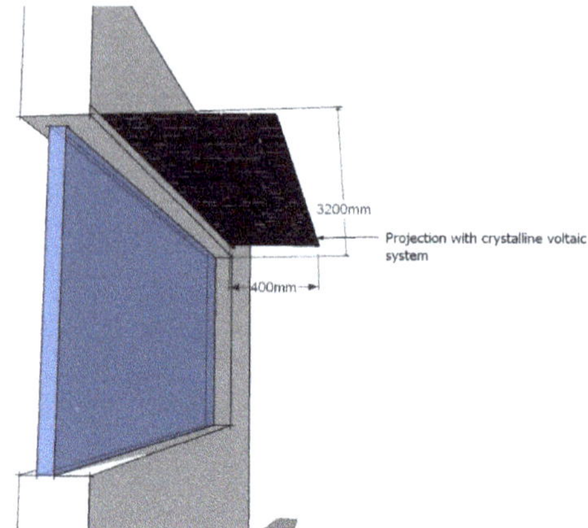

Figure 19. PVSD system used on south façade with mono crystalline photovoltaic system (authors).

Thus, the number of watts decreases. Also, according to the global formula calculation of the solar PV energy output of a photovoltaic system [62]:

$$E = A \times r \times H \times PR \tag{3}$$

where

E = energy (kWh),
A = total solar panel area (20.4 m^2),
r = solar panel yield (estimated to be 12%),
H = annual average irradiation on tilted panels (1800), and PR = performance ratio, coefficient for losses (0.75). The energy becomes more than 3300 kWh/annum.

6. Conclusions

Openings and proper orientation for buildings within the Mediterranean region have always played a critical role in enhancing the comfort level of users. Nevertheless, it comes with a price; it often leads to overheating of the interior space in summer or inadequate penetration of sunlight in winter, especially when the building lacks proper orientation. The result is high energy consumption for heating and cooling. Therefore, developing good strategies that will conserve energy as well as generate clean energy for buildings in Cyprus and other countries facing the Mediterranean Sea is critical. Today, there are several strategies as well as new technology/products that have been designed to enhance the architectural integration of PVs in buildings. Openings on the south façade are often considered most appropriate for integration. This research provides strategies for increasing the comfort level in buildings through the integration of PVs into shading elements in the residential building. Having conducted an empirical study on the use of PVs as a shading device through a simulation on a typical single-family house, this study proposes the use of photovoltaic integrated shading instead of reinforced concrete, which is the commonly used building material for shading devices.

Based on the aim and objectives of this research work, parameters analyzed, and the building simulation, the following conclusions have been obtained:

The simulation result derived from the single-family detached house in Famagusta indicates that the strategic use of PVSDs for openings oriented towards the east, west, and south can reduce its energy consumption by almost 50% in three peak months of the year.

The integration of PVSDs cut down up to 400 kWh of energy consumption through the year and raises the comfort level of the building by about 20%.

PVSDs used as a shading device inclined at $0°$ will provide nearly 2800 watts that can provide up to 50% of the electricity demand of the single-family building.

The cost of the substituted materials and greenfield that would otherwise have been used to mount the panels will be saved.

The authors recommend the use of horizontal overhang on the south façade, while on the east and west façade, overhanded and fins are most appropriate.

Author Contributions: Conceptualization, J.E.O. and E.H.; Methodology, J.E.O. and E.H.; Software, J.E.O.; Validation, E.H., J.E.O.; Formal analysis, J.E.O.; Investigation, E.H.; Resources, J.O., E.H.; Data curation, J.E.O.; Writing—original draft preparation, J.E.O. and E.H.; Writing—review and editing, J.E.O. and E.H.; Visualization, J.E.O.; Supervision, E.H.; Project administration, E.H. and J.E.O.

Funding: This research received no external funding.

Acknowledgments: The authors gratefully acknowledge Wezha Hawez Baiz for her invaluable contributions to the success of this paper.

Conflicts of Interest: The authors declare no conflict of interest.

References

1. Statistical Review of World Energy (June 2016). Available online: https://www.bp.com/en/global/corporate/energy-economics/statistical-review-of-world-energy.html (accessed on 10 November 2018).
2. KIB-TEK (2018). Available online: https://www.kibtek.com/ (accessed on 5 December 2018).
3. Elinwa, U.K.; Radmehr, M.; Ogbeba, J.E. Alternative energy solutions using BIPV in apartment buildings of developing countries: A case study of North Cyprus. *Sustainability* **2017**, *9*, 1414. [CrossRef]
4. Ozbalta, T.G. *Fotovoltaik teknolojisi ile bina kabuğunun degisen işlevleri ve yüzeyleri*; Dizayn Konstrüksiyon Issue 281; Anadolu Üniversitesi Mimarlık Fakultesi: Eskişehir, Turkey, 2009; pp. 75–81.
5. Jelle, B.P.; Breivik, C.; Røkenes, H.D. Building integrated photovoltaic products: A state-of-the-art review and future research opportunities. *Sol. Energy Mater. Sol. Cells* **2012**, *100*, 69–96. [CrossRef]
6. Ali, M.M.; Armstrong, P.J. Overview of sustainable design factors in high-rise buildings. In *Proc. of the CTBUH 8th World Congress*; CTBUH: Chicago, IL, USA, 2008; pp. 3–5.
7. Altin, M. Binaların Enerji Ihtiyacinin PV Bilesenli Cepheler ile Azaltilmasi 3. Ulusal Cati ve Cephe Kaplamalarinda Cagdas Malzeme ve Teknolojileri Sempozyumu. 17–18 Ekim 2006, ITU Mimarlık Fakultesi, Taokiola. Retrieved from World Wide Web on 1 June 2011. Available online: http://www.catider.org.tr/pdf/sempozyum/Bil7.pdf (accessed on 10 October 2018).
8. Das, S.K.; Verma, D.; Nema, S.; Nema, R.K. Shading mitigation techniques: State-of-the-art in photovoltaic applications. *Renew. Sustain. Energy Rev.* **2017**, *78*, 369–390. [CrossRef]
9. Garde, F.; Bastide, A.; Wurtz, E.; Achard, G.; Dobre, O.; Thellier, F.; Ottenwelter, E.; Ferjani, N.; Bornarel, A. *ENERPOS: A National French Research Program for Developing New Methods for the Design of Zero Energy Buildings, CESB 07*; Central Europe towards Sustainable Buildings: Prague, Czech Republic, 2007.
10. Li, X.; Strezov, V. Energy and Greenhouse Gas Emission Assessment of Conventional and Solar Assisted Air Conditioning Systems. *Sustainability* **2015**, *7*, 14710–14728. [CrossRef]
11. Zhai, X.Q.; Wang, R.Z.; Dai, Y.J.; Wu, J.Y.; Xu, Y.X.; Ma, Q. Solar integrated energy system for a green building. *Energy Build.* **2007**, *39*, 985–993. [CrossRef]
12. Imal, M.; Yılmaz, K.; Pınarbaşı, A. Energy Efficiency Evaluation and Economic Feasibility Analysis of a Geothermal Heating and Cooling System with a Vapor-Compression Chiller System. *Sustainability* **2015**, *7*, 12926–12946. [CrossRef]
13. Nocera, F.; Faro, A.L.; Costanzo, V.; Raciti, C. Daylight Performance of Classrooms in a Mediterranean School Heritage Building. *Sustainability* **2018**, *10*, 3705. [CrossRef]
14. Costanzo, V.; Evola, G.; Marletta, L.; Nascone, F.P. Application of Climate Based Daylight Modelling to the Refurbishment of a School Building in Sicily. *Sustainability* **2018**, *10*, 2653. [CrossRef]
15. Dunn, J.R.; Hayes, M.V. Social inequality, population health, and housing: A study of two Vancouver neighborhoods. *Soc. Sci. Med.* **2000**, *51*, 563–587. [CrossRef]
16. Sadalla, E.K.; Sheets, V.L. Symbolism in building materials: Self-presentational and cognitive components. *Environ. Behav.* **1993**, *25*, 155–180. [CrossRef]

17. Halpern, D. *Mental Health and the Built Environment: More Than Bricks and Mortar?* Routledge: Abingdon, UK, 2014.

18. Liu, Z.; Osmani, M.; Demian, P.; Baldwin, A.N. *The Potential Use of BIM to Aid Construction Waste Minimalization*; Centre Scientifique et Technique du Bâtiment: Marne-la-Vallée, France, 2011.

19. Cardona, A.J.A.; Chica, C.A.P.; Barragán, D.H.O. *Building-Integrated Photovoltaic Systems (BIPVS): Performance and Modeling Under Outdoor Conditions*; Springer: Berlin, Germany, 2018.

20. Assoa, Y.B.; Mongibello, L.; Carr, A.; Kubicek, B.; Machado, M.; Merten, J.; Misara, S.; Roca, F.; Sprenger, W.; Wagner, M.; et al. Thermal analysis of a BIPV system by various modelling approaches. *Sol. Energy* **2017**, *155*, 1289–1299. [CrossRef]

21. Shukla, A.K.; Sudhakar, K.; Baredar, P.; Mamat, R. Solar PV and BIPV system: Barrier, challenges and policy recommendation in India. *Renew. Sustain. Energy Rev.* **2017**, *82*, 3314–3322. [CrossRef]

22. Heinstein, P.; Ballif, C.; Perret-Aebi, L. Building integrated photovoltaics (BIPV): Review, potentials, barriers and myths. *Green* **2013**, *3*, 125–156. [CrossRef]

23. Dehra, H. An investigation on energy performance assessment of a photovoltaic solar wall under buoyancy-induced and fan-assisted ventilation system. *Appl. Energy* **2017**, *191*, 55–74. [CrossRef]

24. Luo, Y.; Zhang, L.; Liu, Z.; Wu, J.; Zhang, Y.; Wu, Z.; He, X. Performance analysis of a self-adaptive building integrated photovoltaic thermoelectric wall system in hot summer and cold winter zone of China. *Energy* **2017**, *140*, 584–600. [CrossRef]

25. Martin, J., II. BIPV: Building-Integrated Photovoltaics, the Future of PV. 2011. Available online: http://www.solarchoice.net.au/blog/bipv-building-integrated-photovoltaics-the-future-of-pv/ (accessed on 12 October 2016).

26. Hagemann, I. Shading systems with PV A new market for prefabricated building elements? In *Environmentally Friendly Cities: Proceedings of Plea 1998, Passive and Low Energy Architecture, 1998, Lisbon, Portugal, June 1998*; James & James (Science Publishers) Ltd: London, UK, 2014; Volume 373.

27. Tablada, A.; Kosorić, V.; Huang, H.; Chaplin, I.; Lau, S.; Yuan, C.; Lau, S. Design Optimization of Productive Façades: Integrating Photovoltaic and Farming Systems at the Tropical Technologies Laboratory. *Sustainability* **2018**, *10*, 3762. [CrossRef]

28. Kahoorzadeh, A. Improvement of Thermal Comfort in Residential Buildings by Passive Solar Strategies Using Direct Gain Techniques. Ph.D. Thesis, Eastern Mediterranean University (EMU)-Doğu Akdeniz Üniversitesi (DAÜ), Famagusta, Norther Cyprus, 2013.

29. Pester, S.; Crick, F. *Performance of Photovoltaic Systems on Non-Domestic Buildings*; IHS BRE Press: St Albans, UK, 2013.

30. Taveres-Cachat, E.; Bøe, K.; Lobaccaro, G.; Goia, F.; Grynning, S. Balancing competing parameters in search of optimal configurations for a fix louvre blade system with integrated PV. *Energy Procedia* **2017**, *122*, 607–612. [CrossRef]

31. Zhang, W.; Lu, L.; Peng, J. Evaluation of potential benefits of solar photovoltaic shadings in Hong Kong. *Energy* **2017**, *137*, 1152–1158. [CrossRef]

32. Yoo, S.-H.; Lee, E. Efficiency characteristic of building integrated photovoltaics as a shading device. *Build. Environ.* **2002**, *37*, 615–623. [CrossRef]

33. Bahr, W. A comprehensive assessment methodology of the building integrated photovoltaic blind system. *Energy Build.* **2014**, *82*, 703–708. [CrossRef]

34. Nagy, Z.; Svetozarevic, B.; Jayathissa, P.; Begle, M.; Hofer, J.; Lydon, G.; Willmann, A.; Schlueter, A. The adaptive solar facade: From concept to prototypes. *Front. Archit. Res.* **2016**, *5*, 143–156. [CrossRef]

35. Pérez-Alonso, J.; Pérez-García, M.; Pasamontes-Romera, M.; Callejón-Ferre, A.J. Performance analysis and neural modelling of a greenhouse integrated photovoltaic system. *Renew. Sustain. Energy Rev.* **2012**, *16*, 4675–4685. [CrossRef]

36. Aksugur, E. Potential of passive cooling strategies in Cyprus. In *Housing Research Conference, European Network in Housing Research*; ENHR: Helsingor, Denmark, 1996; Volume 174.

37. Oktay, D. Design with the climate in housing environments: An analysis in Northern Cyprus. *Build. Environ.* **2002**, *37*, 1003–1012. [CrossRef]

38. iSHAN JAIN (2017) How to Design A Shading Device? Available online: http://talkarchitecture.in/how-to-design-a-shading-device/ (accessed on 7 August 2018).

39. Kuhn, T.E.; Bühler, C.; Platzer, W.J. Evaluation of overheating protection with sun-shading systems. *Sol. Energy* **2001**, *69*, 59–74. [CrossRef]
40. Valladares-Rendón, L.G.; Schmid, G.; Lo, S. Review on energy savings by solar control techniques and optimal building orientation for the strategic placement of façade shading systems. *Energy Build.* **2017**, *140*, 458–479. [CrossRef]
41. Kirimtat, A.; Koyunbaba, B.K.; Chatzikonstantinou, I.; Sariyildiz, S. Review of simulation modeling for shading devices in buildings. *Renew. Sustain. Energy Rev.* **2016**, *53*, 23–49. [CrossRef]
42. Zhang, X.; Lau, S.; Lau, S.S.Y.; Zhao, Y. Photovoltaic integrated shading devices (PVSDs): A review. *Sol. Energy* **2018**, *170*, 947–968. [CrossRef]
43. James, T.; Goodrich, A.; Woodhouse, M.; Margolis, R.; Ong, S. Building-Integrated Photovoltaics (BIPV) in the residential sector: An analysis of installed rooftop system prices. *Contract* **2011**, *303*, 275–3000.
44. Weller, B. *Detail Practice: Photovoltaics*; Birkhäuser: Basel, Switzerland, 2010.
45. Leskovar, V.Ž.; Premrov, M. An approach in architectural design of energy-efficient timber buildings with a focus on the optimal glazing size in the south-oriented façade. *Energy Build.* **2011**, *43*, 3410–3418. [CrossRef]
46. Boxwell, M. *The Solar Electricity Handbook-2017 Edition: A Simple, Practical Guide to Solar Energy–Designing and Installing Solar Photovoltaic Systems*; Greenstream Publishing: Coventry, UK, 2017.
47. Ozay, N. A comparative study of climatically responsive house design at various periods of Northern Cyprus architecture. *Build. Environ.* **2005**, *40*, 841–852. [CrossRef]
48. TRNC (2006) TRNC General Population and Housing Unit Census. Available online: http://nufussayimi.devplan.org/Census%202006.pdf (accessed on 5 December 2018).
49. Özdeniz, M.B.; Hançer, P. Suitable roof constructions for warm climates—Gazimağusa case. *Energy Build.* **2005**, *37*, 643–649. [CrossRef]
50. Hancer, P. Thermal Insulations of Roofs for Warm Climates. Ph.D. Thesis, Eastern Mediterranean University, Famagusta, Northern Cyprus, 2005.
51. ontheworldmap.com. Available online: http://ontheworldmap.com/cyprus/city/famagusta/famagusta-location-on-the-cyprus-map.jpg (accessed on 15 January 2019).
52. Lapithis, P.A. *Passive solar architecture in Cyprus*; Lapithis Architectural Firm: Lefkosa, Cyprus, 2005.
53. Climatemps.com. Available online: http://www.famagusta.climatemps.com/ (accessed on 14 August 2018).
54. Pourvahidi, P. Bioclimatic Analysis of Vernacular Iranian Architecture. Ph.D. Thesis, Eastern Mediterranean University (EMU), Famagusta, Northern Cyprus, 2010.
55. Atalar, E. Evaluation of Solar Insolation in Northern Cyprus. Ph.D. Thesis, Eastern Mediterranean University, Famagusta, Northern Cyprus, 2001.
56. Theodoros, Z.; Hadjikyriakou, C. State of the Art of Power Generation in Cyprus. In *Social Costs and Benefits of Renewable Electricity Generation in Cyprus*; Springer: Cham, Switzerland, 2016; pp. 7–16.
57. Solar resource map © 2018 Solargi. Available online: https://solargis.com/maps-and-gis-data/download/cyprus (accessed on 21 January 2019).
58. Energyplus (2018), EnergyPlus. Available online: https://energyplus.net/ (accessed on 4 December 2018).
59. Reta, T. Leveraging a Building Information Model to Carry Out Building Energy Performance Analysis. Bachelor's Thesis, Helsinki Metropolia University of Applied Sciences, Helsinki, Finland, 2017.
60. Murray, M.; Liggett, R.; Benson, A.; Bhattacharya, Y. Climate Consultant 4.0 develops design guidelines for each unique climate. In *American Solar Energy Society Meeting*; ASES: Boulder, CO, USA, 2009.
61. Mesda, Y. Heat Transfer Coefficient Analysis of the Chamber of Cyprus Turkish Architects Office Building on the Zahra Street in the Walled City of Nicosia in Cyprus. *Archit. Res.* **2012**, *2*, 47–54. [CrossRef]
62. Optimum Tilt of Solar Panels. Available online: https://www.solarpaneltilt.com/ (accessed on 28 December 2018).

Article

Parametric Optimization of Window-to-Wall Ratio for Passive Buildings Adopting A Scripting Methodology to Dynamic-Energy Simulation

Giacomo Chiesa [1,*]**, Andrea Acquaviva** [2]**, Mario Grosso** [1]**, Lorenzo Bottaccioli** [3]**, Maurizio Floridia** [4]**, Edoardo Pristeri** [4] **and Edoardo Maria Sanna** [4]

[1]   Department of Architecture and Design, Politecnico di Torino, Turin 10125, Italy; mario.grosso@polito.it
[2]   Department DIST, Politecnico di Torino, Turin 10125, Italy; andrea.acquaviva@polito.it
[3]   Department DAUIN, Politecnico di Torino, Turin 10138, Italy; lorenzo.bottaccioli@polito.it
[4]   ICT for Smart Societies, Department DET, Politecnico di Torino, Turin 10138, Italy;
     maurizio.floridia@studenti.polito.it (M.F.); edoardo.pristeri@studenti.polito.it (E.P.);
     edoardomaria.sanna@studenti.polito.it (E.M.S.)
*   Correspondence: giacomo.chiesa@polito.it; Tel.: +39-011-090-4376

Received: 16 May 2019; Accepted: 28 May 2019; Published: 31 May 2019

Abstract: Counterbalancing climate change is one of the biggest challenges for engineers around the world. One of the areas in which optimization techniques can be used to reduce energy needs, and with that the pollution derived from its production, is building design. With this study of a generic office located both in a northern country and in a temperate/Mediterranean site, we want to introduce a coding approach to dynamic energy simulation, able to suggest, from the early-design phases when the main building forms are defined, optimal configurations considering the energy needs for heating, cooling and lighting. Generally, early-design considerations of energy need reduction focus on the winter season only, in line with the current regulations; nevertheless a more holistic approach is needed to include other high consumption voices, e.g., for space cooling and lighting. The main considered design parameter is the WWR (window-to-wall ratio), even if further variables are considered in a set of parallel analyses (level of insulation, orientation, activation of low-cooling strategies including shading devices and ventilative cooling). Finally, the effect of different levels of occupancy was included in the analysis to regress results and compare the WWR with corresponding heating and cooling needs. This approach is adapted to Passivhaus design optimization, working on energy need minimisation acting on envelope design choices. The results demonstrate that it is essential to include, from the early-design configurations, a larger set of variables in order to optimize the expected energy needs on the basis of different aspects (cooling, heating, lighting, design choices). Coding is performed using Python scripting, while dynamic energy simulations are based on EnergyPlus.

Keywords: environmental and technological design; passive cooling systems; energy need optimisation; passivhaus; massive simulation modelling; regression analysis

---

1. Introduction

Buildings are responsible for more than 40% of the total primary energy consumption in industrialized countries [1–4], and roughly one third of the relevant GHG emissions. Considering this great influence of the building sector on national energy balances, several actions have been taken by government institutions in order to: firstly, reduce the building energy needs; secondly, increase the efficiency of the installed equipment; and thirdly, increase the amount of energy produced by renewable sources. At the European level, the EPBD directive and further upgrading—see the

EPBD recast and the recent Directive 2018/844—have progressively acted on the reduction of energy needs and consumptions, while other directives, such as the 2009/28/EC, have worked to promote the usage of renewable sources. Nevertheless, while much attention was paid to reducing heating energy consumption, the reduction of cooling energy needs has not elicited the same consideration. Cooling energy needs have been constantly growing due to several causes, including climate change, international building styles, and changes in the culture of comfort [5–7]. Furthermore, the adoption of extended insulation levels may cause an increase in overheating effects—see for example [8,9]. The need to include in the design process low-energy cooling strategies in order to correctly balance energy needs was also underlined by several authors [10,11].

1.1. WWR and energy needs – a short background analysis

Design optimization studies, considering both winter and summer effects, are essential to avoid the adoption of flawed design decisions from the sustainable/environmental point of view. The impact of envelope design choices on energy needs is evident. Among several design aspects related to envelope definition, the ratio of transparent and opaque areas, i.e. the WWR (Window-to-Wall Ratio), is recognized to have a high impact on building energy balances [12]. Since the 1970s, studies have been conducted to define optimal WWR values corresponding to the minimal annual energy needs [13,14]. Nevertheless, these first analyses do not include the effect of passive cooling solutions (e.g., shading systems or CNV) nor the impact of internal gain (occupancy) variations. The relationship between WWR and energy needs was also studied in the 1990s within the EC-funded Project LT (lighting thermal), wherein the influence of WWR on lighting, space cooling and heating in buildings, for average southern European climate conditions, was analysed. Results showed that the minimum yearly-balanced energy consumption in a residential building could be found at WWR values of: 8%, 10%, 15%, respectively for horizontal, East/West, and South/North window exposure, if windows are not shaded; 10%, 15%, 20%, and 30%, respectively, for horizontal, North, East/West, and South window exposure, with 65% of windows shaded [15]. However, this analysis was based on a simplified calculation method and did not include the effect of occupancy nor the impact of passive cooling solutions (e.g., CNV), nor the effect of different levels of insulation.

In 2010, the relation between WWR and thermal energy needs, was studied for a large office building in Shanghai, China, including life cycle assessment results. Nevertheless, only thermal results were included based on a spreadsheet calculation. In total, 63 cases were simulated showing a positive correlation between an increase in the WWR and an environmental impact reduction [16]. Also, in this case, passive cooling solutions and internal gains were not considered, while the effect of WWR on lighting was also not investigated. Another approach was presented in [17], focusing on the effect that climate indicators, i.e. the ambient temperature amplitude, and envelope U-value have on the definition of the maximum WWR for reaching thermal autonomy in buildings. 135 simulations were carried out considering seven U.S. locations. For this analysis no HVAC systems were included considering free-running operation. Results underline the need to define methodologies able to suggest WWR values from the early-design approaches to consider local climates and different envelope thermal transmittances. Authors further suggest the usage of statistical correlations in further studies. No passive cooling solutions, nor internal gains or lighting needs are considered in this analysis.

Furthermore, geographical studies on optimal WWR definition were conducted in Ref. [18], considering five Asian locations, and in Ref. [19], focusing on four locations, two in U.S. and 2 in Europe. The first study investigated the relation between WWR and total energy performance, while the second focused on energy needs for heating and cooling. In both cases no passive cooling solutions or internal gain variations were assumed. Differently, a detailed analysis on optimal WWR definition in relation to energy needs for heating, cooling and lighting was performed for four European locations [20]. This paper includes the effect of shading systems by also considering different activation flux thresholds. A low energy office building was simulated in EnergyPlus considering one U-value configuration. Internal loads were defined in compliance to standard values. A sensitivity analysis was also conducted considering building compactness, equipment efficiency and artificial light efficiency. In addition, natural light analyses were performed considering the UDI and the DA indexes. The investigation is based on five WWR in order to define potential correlation curves. Nevertheless, this study does not include CNV, random internal gain variations, or the impact of different U-value on results.

Other recent studies [21,22] have also defined potential optimization levels for WWR. Interesting graphical models for designers were produced referring to a middle European case study of a sample shaped building simulated in EnergyPlus with average levels of insulation and no internal gains. These graphs include cooling and heating energy needs for different orientations in the range E-S-W, different building shapes and three WWR values [21]. Furthermore, an optimisation analysis to define some envelope characteristics for a sample building within rural and urban contexts was also produced in [22] considering dynamic energy simulations including heating, cooling and lighting energy needs. In particular, this study refers to the number, dimension, position of windows and wall thickness. A parametric investigation of Italian conditions was carried out in 2017 [23] considering 12 locations, different U-values, i.e., low and high insulation, and for the latter, normal and spectral selective glazing cases, and seven WWR steps for a total of 518 simulations. In this analysis, the shading effect was also included considering electrochromic glazing, but not the effect of CNV or internal gain variations. The authors underlined the high effect of climate on optimal WWR, while other aspects did not seem to vary considerably this parameter, even if they suggested to analyse them more in details. In the same year in Ref. [24], the relation between WWR and window orientation for an office building localize in Tripoli, Libya, was investigated. The considered case study is simplified by a schematic box with one non adiabatic wall confining with the external environment. Eight orientations and 10 WWR steps were considered for this analysis conducted via EnergyPlus. The analysis focuses on cooling and heating loads, while other aspects are fixed in accordance to ASHRAE suggestions. Results showed a direct correlation between annual energy needs (cooling and heating) and WWR in all orientations, even if this effect was higher for southern cases (SE, S, SW, W). This study did not include the effect of natural light balance, passive cooling solutions (nor shading or CNV), nor internal gain variations or the influence of different U-values or climates, even if some of these aspects are expected to be included in future developments.

A method to map the suggested WWRs was investigated in [25] for 10 Japan locations. Three classes were defined: (i) WWR directly related to CO_2 emission, (ii) optimal WWR minimizing CO_2 can be defined, (iii) WWR and CO_2 are inversely correlated. A typical office building was used as reference case study to perform EnergyPlus dynamic energy simulations. Results included cooling, heating and lighting energy needs, considering they transposition in equivalent CO_2 emissions [kg]. Four orientations, two lighting powers (5 and 10 W/m^2), and seven WWRs were included in the proposed approach. Fixed thermal properties were assumed by national standards. This paper also investigated the effect of three internal gain levels by varying together occupancy and equipment densities. Results showed that internal gains principally effected CO_2 emissions levels, even if, in some cases, they also influence the optimal WWR. Results suggested that this aspect may be further investigated. Nevertheless, the effect of passive cooling strategies or different thermal envelope characteristics were not included. Furthermore, the correlation between NZEB buildings and WWR was investigated in [26], considering a severely cold China location (Shenyang). A simple building was simulated in EnergyPlus to define heating and cooling energy needs in accordance with different WWRs and three orientations. Considering the rigid climate condition, a direct relation between the WWR and the energy needs was underlined for all orientations. Fixed thermal characteristics of the building were adopted. No passive cooling solutions, U-value, internal gain variations, daylight balance, or regression analyses were considered.

In 2019 the correlation between energy and daylight performance of a sample office room south-oriented for different WWRs was tested considering different percentage of integration of CdTe PV gazing in windows [27]. Five locations, representative of each Chinese climate zone, were selected. A total of 28 cases were simulated for each of them, considering 4 WWR steps and different PV integration percentages. Results showed that PV windows can help in reducing energy needs in office buildings starting from large WWR, ≥45%. This study is based on EnergyPlus and Radiance. No thermal characteristics or internal gain variations are considered. Furthermore, the effect of passive cooling solutions was not investigated. Finally, an optimization analysis of WWR in China low latitude region were presented in 2019 [28] considering also the effect of fixed external sunshade systems (overhang, vertical and comprehensive cases). Cooling, heating and lighting energy needs were considered in this optimization analysis. A sample hotel building was assumed as a reference to perform the simulations in EnergyPlus and Radiance considering four orientations. Results from both software programs were used to optimize the WWR. These analyses aimed to correlate the minimal WWR to reach daylight standards with energy consumptions.

The present paper focuses on the influence that specific façade design choices have on the expected building energy needs for heating, cooling and lighting in the preliminary phase, when the possibility to change is higher and its cost lower, assuming an environmental and technological approach—see [29]. This approach is consistent with the "passive house" concept as developed by the Passive House Institute of Darmstad, Germany [30]. In particular, the presented analysis deals with the influence of WWR on the energy need of an office building, in combination to other parameters such as envelope Uvalue [31], windows orientation, shading coefficient, and controlled natural ventilation (CNV) to perform ventilative cooling, i.e., wind-driven and stack-driven airflow through openings controlled by motorised actuators linked to microclimate sensors. The dynamic energy simulations were carried out using EnergyPlus and Python for two reference locations. The proposed investigation not only analyses the obtained results considering the proposed case studies, but is based on the elaboration of a code that can be used to model the optimal WWR in different locations or for different configurations. Furthermore, the reliability of results was checked under the influence of random internal gain variations (occupancy level) in order to define the statistical correlation curves. The adoption of a scripting simulation approach allows, in fact, to increase the number of performed simulations by two or three orders of magnitude with respect to previous analyses. Thanks to the inclusion of all these aspects, the followed approach can be considered innovative in comparison to previous research on the topic.

1.2. The Research Objective and Structure

The main objective of this study was to develop an algorithm to optimise, from the early-design phase, the WWR of an office building for reducing the expected energy needs for space heating and cooling, and lighting to the levels required by the Passive House concept. This study was conducted following a multidisciplinary approach, based on the collaboration between ICT Master Degree students and researchers from different fields: architectural technology and environmental design, telecommunication engineering, and data elaboration. The proposed approach is suited to the design of new constructions as well as major building refurbishments, and it is applied here to two locations: one, Helsinki, with a harsh winter climate; the other, Turin, with a temperate climate, located in the enlarged Mediterranean area. Of course, the methodology of this paper can be applied to different climates in order to demonstrate how design optimisation choices differs according to environmental conditions.

In addition to WWR, other parameters were considered in the energy optimisation analysis: envelope thermal transmission (opaque and transparent); CNV; and window shading coefficient. Two different European locations, Helsinki and Turin, representing, respectively, cold and temperate climate conditions, were considered. Moreover, the effect of random changes in the occupancy level was included in the analysis to improve the resilience of the proposed optimisation models in relation to the impact of internal gains variations (i.e., the presence of people).

The paper is structured as follows: in Section 2 the proposed methodology is introduced; Section 3 describes the results of the WWR optimisation process; Section 4 is related to the discussion of results including the effect of random occupancy and monthly evaluation of heating, cooling and lighting energy needs; paper's conclusions are described in Section 5.

2. Materials and Methods

The proposed approach is based on the definition of a script to generate parametric analysis outputs, e.g., graphs, showing the energy needs for the heating, cooling and lighting of a sample office unit as function of WWR. The simulation programmes used are: Design Builder v5.5 (DB) (DesignBuilder Software Ltd, Gloucs, UK), to generate the starting case study; EnergyPlus, to perform dynamic energy simulations; and Python, to control the whole process, modify input data, collect simulation results, and analyse output data including graph elaborations.

DB was used to create the 3D model of the reference office-building unit, described in Section 2.1., while EnergyPlus was used—via a Python script—to simulate its energy needs for various envelope configurations, and for generating the relevant *.idf files. The Python library Geomeppy, allowed for changing parametrically WWR as well as running directly simulations, without using the EnergyPlus interface each time, was adopted.

Furthermore, a sensitivity analysis, able to consider the potential variation effect in energy needs due to random variations in the internal loads, was carried out. These variations were based on the occupancy levels, simulating real operation, without adding equipment, which will be the topic of a future improvement of the method. In each simulation, the standard occupancy datum defined in DB —standard office schedule—was let varying randomly according to a Gaussian distribution—$G(\mu,\sigma)$, with $\mu = 0.09225$ people/m^2 and $\sigma = 0.14075$ people/m^2—in order to make a more realistic impact on the energy needs of the internal-gain variations due to the presence of people. The variation domain was adapted by [32]. A sample plot of 10,000 random values extracted by the adopted Gaussian is reported in Figure 1 in order to show, considering the central limit theorem, that the chosen values are statistically reasonable. The values reported in this figure refer to the net simulated office area of 80 m^2.

Simulation outputs include heating and cooling, as well as annual lighting energy needs. The light requirement was set to 400 lux at desk height, balancing illuminance requirements on task and surrounding areas for offices according to UNI EN 12464-1 while the linear dimmer control was assumed in EnergyPlus to balance the positive effect of natural lighting with the additional need for artificial sources. Internal normalised light loads were assumed to be 5 W/m^2-100lux. This value is of course balanced by the software in accordance to the amount of natural light calculated by EnergyPlus. The use of this dynamic simulation tool to also simulate natural/artificial balance was demonstrated to be effective by several authors, see for example the discussion in [20]. Heating and cooling set points were assumed to be respectively 20 °C and 26 °C.

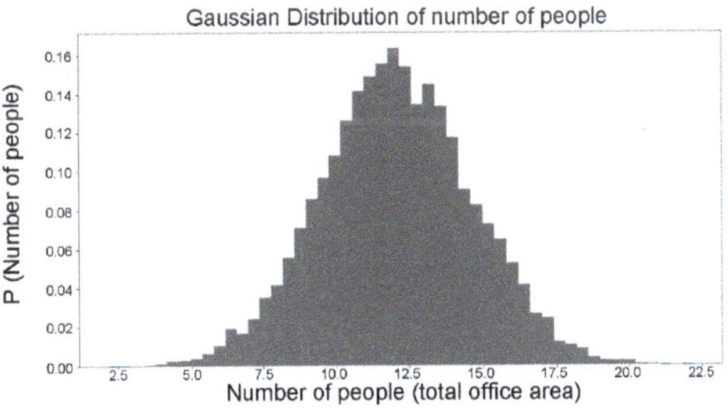

Figure 1. Sample plot of 10,000 random occupancy values generated by the used Gaussian model – total office area = 80 m^2.

The analysis was carried out according to the following phases.

1. A simulation with occupancy level constant during the year and varying WWR from 1% to 95%, by a step of 5%.
2. Simulations with variation of the occupancy level within a 16-values range for each configuration, based on the above-mentioned Gaussian distribution; and 16128 runs for each location, considering the 21 WWR variations and the 48 building configurations described in Section 2.1. This step aimed at creating train and test datasets for statistical analyses.
3. Regression analyses of the output data, divided in train—to develop the regression—and test sets by a ratio 70–30%, as well as a calculation of the RMSE of the regression with respect to the independent test database.

Figure 2 shows a flowchart of the developed simulation engine.

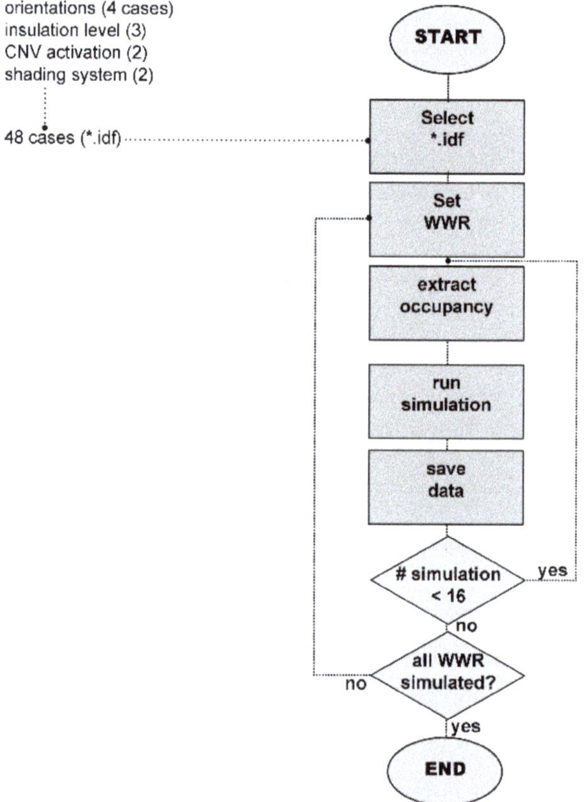

Figure 2. Sketch of the developed simulation engine.

2.1. The Case Study

The case study is an office space unit virtually included in a multi-storey building as shown in Figure 3. The walls adjacent to the other space units and the floor are assumed as adiabatic, while one wall and the roof are external surfaces. The net area of the space unit was set to 80 m^2 considering an open office space, while the infiltration rate was fixed to 0.7 ACH

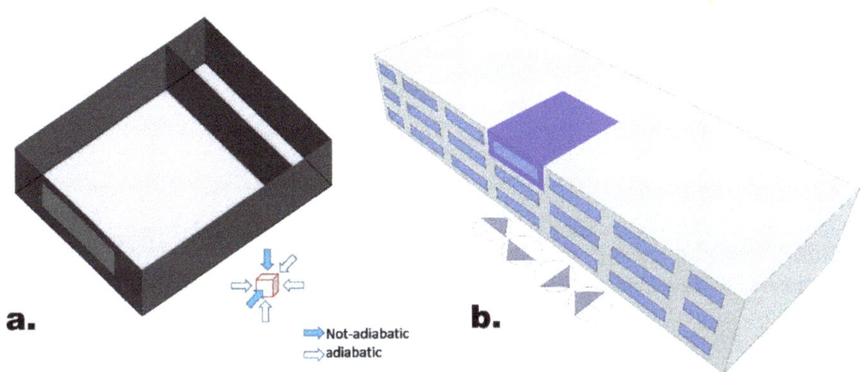

Figure 3. Visualization of (**a**) the reference office unit (open office room + corridor), and (**b**) its virtual position in a multi-storey building.

The configurations applied in simulation are the following:

- Two locations, i.e., Helsinki (FIN) and Turin (ITA);
- Three values of the envelope heat transmission coefficient (Uvalue) corresponding to low, medium, and high insulation levels (see Table 1 for the opaque envelope, and Table 2 for windows);
- Four orientations of the external wall, i.e., South, East, West, and North;
- Shading devices set according to the integrated shading control system of DB—see below—(present/not present) and;
- CNV set to on and off.

Without considering locations, 48 configurations were assumed.

Figure 4 shows materials and layers of the opaque envelope (external wall and roof) in the three considered configurations.

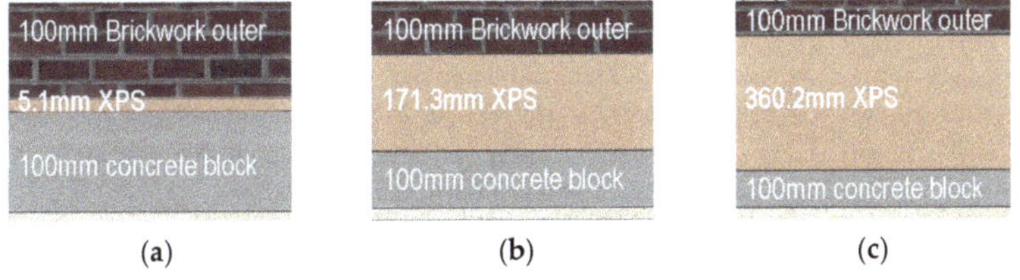

Figure 4. Layers and materials for the 3 configurations of opaque envelope: (**a**) Low insulation (almost null), (**b**) Medium insulation, (**c**) High insulation.

Table 1. Insulation levels of the opaque envelope components (external wall and roof).

Configuration	U-Value Walls [W/m^2]	U-Value Roof [W/m^2]
Low insulation (non-insulated)	1.5	1.5
Medium insulation	0.18	0.18
High insulation	0.09	0.09

Table 2. Insulation levels of windows and relevant materials.

Configuration	Glass Type	U-value Windows [W/m^2]
Low insulation (non-insulated)	Single glazing, clear	5.7
Medium insulation	Double glazing, clear, LoE, argon-filled	1.49
High insulation	Triple glazing, clear, LoE, argon-filled	0.78

The external wall has one window whose dimensions were changed automatically acting on the WWR indicator. In addition, the effects of shading and CNV were considered, together or singularly. In particular, for shading devices an integrated external located blind system with medium reflectivity slats was assumed. The control type for this shading considers both an external air temperature threshold and a solar radiation set point. The first was set to 18 °C, considering the effect of office equipment [33] in comparison to the potential threshold for residential buildings of 21 °C suggested by Olgyay [34], while the second was assumed as 120 W/m^2, in line with the suggested set points—e.g., [33]. Differently, CNV was simulated considering summer activation with a maximal external air control set to 26 °C and a fixed ACH of 6 vol/h in line with ACH values used in other references for early-design [8]. CNV is principally conceived to be naturally-driven, even if small fan-assisted extractors may be activated when buoyancy or wind-driven flows are not sufficient for cooling purposes. In accordance with other ventilative cooling approaches for early-design stages—e.g., early analysis of CNV climate potential by IEA EBC Annex 62—CNV evaluations were performed without including fan energy consumption assuming the prevalent natural-driven force. This is in accordance with the definition of the specific list of input parameters for programmed natural ventilation in EnergyPlus (early-design option). Section 3 describes the simulation results for the 48 combinations assumed for each location.

3. Simulation Results and Analysis

The first analysis step was performed by running all the simulations considering a constant occupancy schedule while changing the WWR. Here, "constant occupancy" means that the only variation of the occupancy happens through the schedule of Design Builder and the mean occupancy value remains constant throughout the year. The schedule used for simulating the occupancy during the day is suited for an open plan office area as provided by the file OpenOff_Occ of DesignBuilder [35].

Figure 5 shows the results of simulations for all configurations and setups, including shading and CNV settings, for the Helsinki case, while Table 3 reports the optimal WWR [%] and related total energy needs [kWh/m^2]. At a first glance, the minimum energy need corresponds to 95% WWR in the high insulated scenario for the South window orientation when both CNV and shading are activated. This is due to the high energy contribution of solar gains in winter, with an optimal control of the potential overheating in summer, as allowed by the South window exposure at the considered high Northern latitude [36,37]. Differently, with the North-facing window the minimum energy need is reached at 55% of WWR with shading system set to off due to the prevailing need for reducing heat losses in winter. East and West window orientations show similar behaviour.

In addition, the higher the WWR, the lower the energy needs for lighting. The above-mentioned behaviours are even more apparent in the cases of "non-insulated" and "medium-insulated" scenarios.

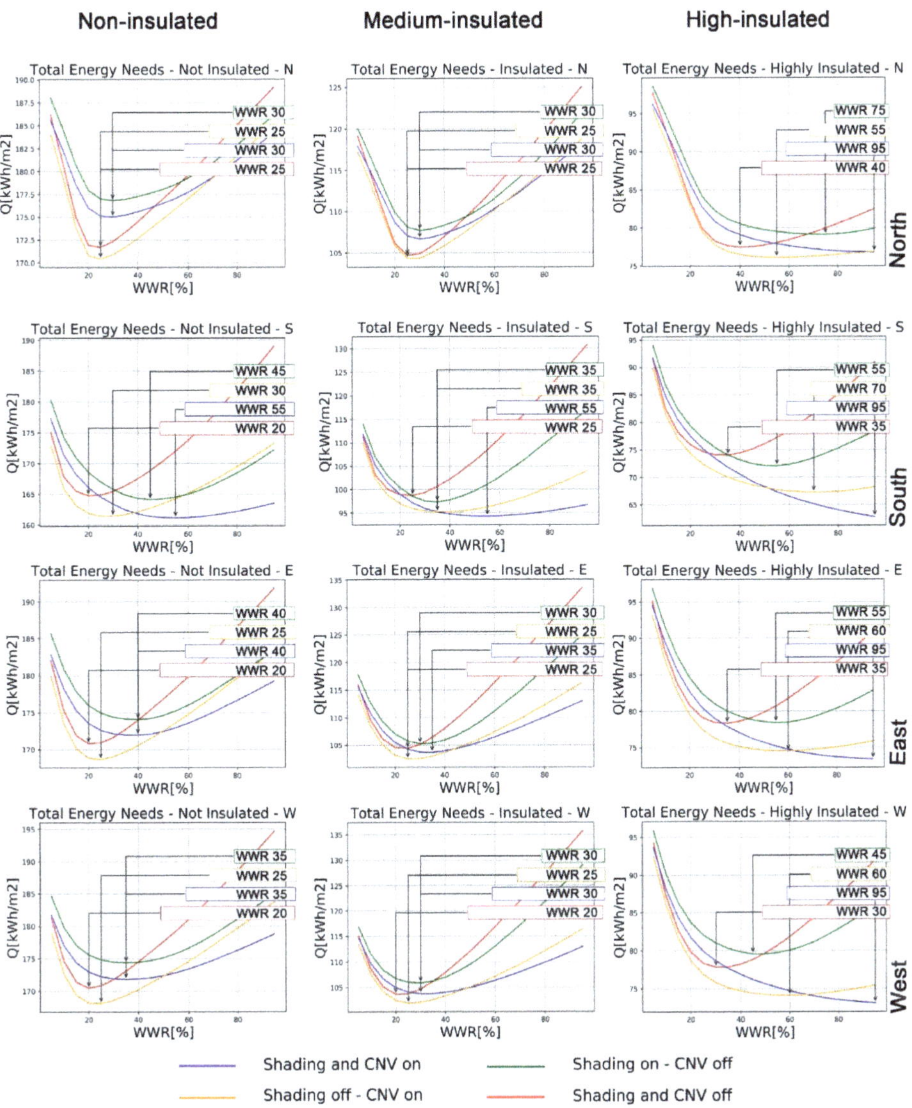

Figure 5. Annual energy need as a function of WWR for various envelope configurations—Helsinki case.

Table 3. Optimal WWR [%] and related total energy needs [kWh/m^2] for various envelope configurations. Helsinki case.

	North		South		East		West	
Non-insulated								
Case	**WWR**	**Q** [kWh/m^2]	**WWR**	**Q** [kWh/m^2]	**WWR**	**Q** [kWh/m^2]	**WWR**	**Q** [kWh/m^2]
Shading and CNV on	30	175.03	55	161.17	40	172.01	35	171.84
Shading on, CNV off	30	176.84	45	164.12	40	174.09	35	174.40
Shading off, CNV on	25	170.45	30	161.45	25	168.62	25	168.12
Shading and CNV off	25	171.72	20	164.78	20	170.84	20	170.51
Medium-insulated								
Case	**WWR**	**Q** [kWh/m^2]	**WWR**	**Q** [kWh/m^2]	**WWR**	**Q** [kWh/m^2]	**WWR**	**Q** [kWh/m^2]
Shading and CNV on	30	106.67	55	94.33	35	103.70	30	103.77
Shading on, CNV off	30	107.71	35	97.34	30	105.28	30	105.88
Shading off, CNV on	25	104.33	35	95.24	25	102.49	25	101.93
Shading and CNV off	25	104.69	25	98.75	25	104.42	20	103.67
High-insulated								
Case	**WWR**	**Q** [kWh/m^2]	**WWR**	**Q** [kWh/m^2]	**WWR**	**Q** [kWh/m^2]	**WWR**	**Q** [kWh/m^2]
Shading and CNV on	95	76.79	95	62.76	95	73.42	95	73.11
Shading on, CNV off	75	79.10	55	72.13	55	78.46	45	79.57
Shading off, CNV on	55	76.1	70	67.26	60	74.55	60	74.12
Shading and CNV off	40	77.47	35	74.07	35	78.36	30	77.84

Similarly, Figure 6 shows the final simulation results on the annual energy needs for heating, cooling, and lighting of the reference office-building unit located in Turin in all envelope configurations, while Table 4 reports optimal WWR and total energy needs for all considered cases. The following comments could be made on these results

- As expected, in the absolute values, the energy needs are always higher with lower insulation levels for any window orientation; the lowest amount of energy needs for each insulation level is reached with a Southern window orientation, whereby it is easier to reduce solar radiation in summer while solar gains contribute to space heating in winter.

- Energy needs decrease with increasing WWR up to a certain %, with changes depending on both window orientation and insulation level.

- If a window is shaded, this trend inversion occurs in the range of 60% to 90%, due to a negative solar gains unbalance between winter and summer, in absence of heat dissipation by CNV; in fact, a shift of the trend inversion towards lower WWR values, and lower energy needs occur in the case of CNV on.

- In the absence of shading, an abrupt decrease of energy needs occurs up to 20–50% of WWR, with an inversion of this trend afterword and always lower values if CNV is on; within the above-mentioned range, the minimum values are shifted towards higher values of WWR in the case of CNV on.

- Considering a WWR around 30% as a common average value in the current building design practice, an optimal window configuration, corresponding to the lowest annual combined energy need, is given by the case with shading off and CNV on for all orientations.

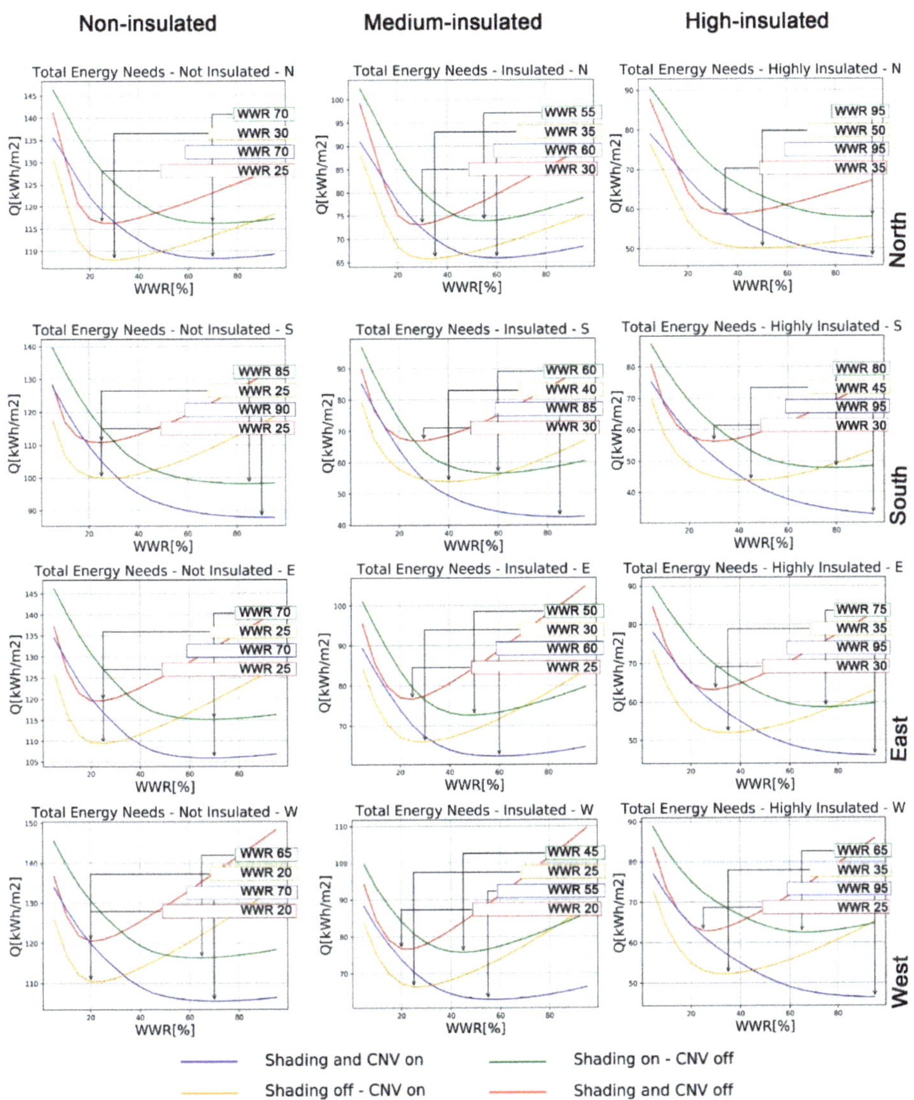

Figure 6. Annual energy need as a function of WWR for various envelope configurations—Turin case.

Table 4. Optimal WWR [%] and related total energy needs [kWh/m^2] for various envelope configurations. Turin case.

| | Non-insulated | | | | | | | |
| | North | | South | | East | | West | |
Case	WWR	Q [kWh/m^2]	WWR	Q [kWh/m^2]	WWR	Q [kWh/m^2]	WWR	Q [kWh/m^2]
Shading and CNV on	70	108.37	90	87.89	70	105.98	70	105.59
Shading on, CNV off	70	116.26	85	98.10	70	115.24	65	116.34
Shading off, CNV on	30	108.05	25	99.95	25	109.49	20	110.51
Shading and CNV off	25	116.33	25	110.83	25	119.70	20	120.52
	Medium-insulated							
	North		South		East		West	
Case	WWR	Q [kWh/m^2]	WWR	Q [kWh/m^2]	WWR	Q [kWh/m^2]	WWR	Q [kWh/m^2]
Shading and CNV on	60	65.98	85	42.55	60	62.29	55	62.93
Shading on, CNV off	55	73.93	60	56.60	50	72.68	45	75.70
Shading off, CNV on	35	65.88	40	53.83	30	65.90	25	66.33
Shading and CNV off	30	73.16	30	66.98	25	76.66	20	76.71
	High-insulated							
	North		South		East		West	
Case	WWR	Q [kWh/m^2]	WWR	Q [kWh/m^2]	WWR	Q [kWh/m^2]	WWR	Q [kWh/m^2]
Shading and CNV on	95	47.83	95	32.99	95	46.12	95	46.49
Shading on, CNV off	95	57.97	80	47.86	75	58.71	65	62.50
Shading off, CNV on	50	50.13	45	43.77	35	52.12	35	52.32
Shading and CNV off	35	58.63	30	56.30	30	63.22	25	62.94

The graphs of Figures 7 and 8 show that the higher the WWR, the lower the lighting energy needs due to increased daylight, while it increases the risk of summer overheating. In winter, the increase in solar gains due to a larger window is counterbalanced, in almost all cases, by an increase of thermal losses, due to the higher-value of glazing in respect to opaque walls.

The graphs (**a**) in Figures 7 and 8 refer to the North position, in which the window receives little or no direct solar radiation, while graphs (**b**) are related to the window exposed towards South, intercepting solar radiation during the hours when it is most energy intensive. Considering the South and North facing windows, the cases of Helsinki and Turin are differentiated, as expected, by higher values of energy needed for heating and much lower values for cooling in the former location, while lighting energy needs are similar in both cases. Heating energy need, which is not dependent of shading and CNV in both locations and orientations, is affected by WWR for the North-oriented window, according to the trend of a continuous increase in Helsinki and up to 60% WWR, with an almost constant trend afterword, in Turin. For the South-oriented window, heating energy need increases up to 25% WWR in Helsinki, and 15% WWR in Turin; above those values, an abrupt change in trend occurs with a continuous decrease. Regarding cooling energy needs, which are negligible in Helsinki for a North-facing window, shading is the most affecting condition with an abrupt change in trend from a decrease to increase of energy need at 10% WWR in both locations for a non-shaded South-oriented window. The addition of shading in the South-oriented window has an effect of keeping the cooling energy need almost constant with WWR in Turin, while increases above 30% WWR in Helsinki. Combining shading and opening for CNV in the South-oriented window, practically zeros cooling energy in Helsinki and lower it to a negligible value in Turin.

Regarding graphs (**c**)—East-oriented window—and (**d**)—West-oriented window—in the case of Helsinki, the trend of heating energy need is similar to the one of the North-facing window, while the trends of cooling energy are similar to the ones of the South-facing window. In Turin, the trend of heating energy need is similar to the one of the South-facing window, but with a shift of the inversion peak to 40% and 35% WWR for East and West orientation, respectively; while the trends of cooling energy need are similar to the ones of the South-facing window.

The energy need for lighting decreases continuously with WWR by a non-linear trend in both locations and it is affected only slightly by the shading condition, more so in Turin than in Helsinki.

Since lighting energy needs are particularly high in office buildings, they have a significant impact on the definition of the optimal WWR configuration. In fact, a reduction of WWR, which could increase energy efficiency if only heating and cooling needs were taken into account, would not have the same effect if considering lighting energy needs as well.

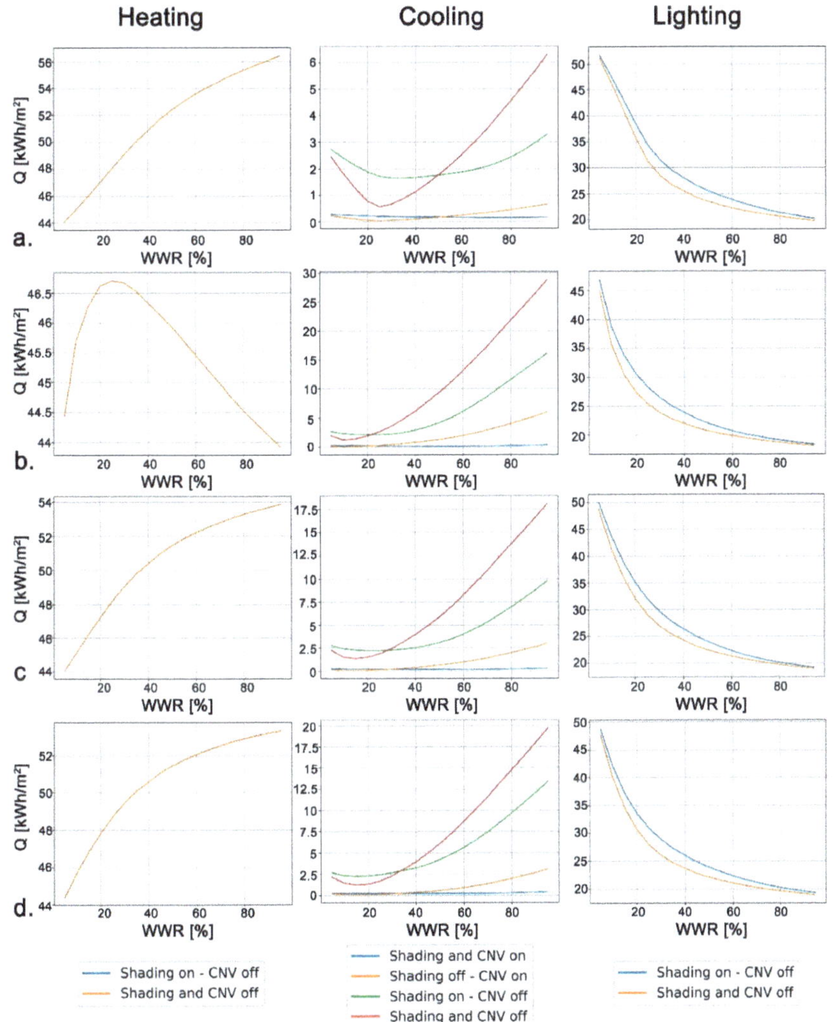

Figure 7. Heating, Cooling, and lighting annual energy needs as a function of WWR for different setting of shading and CNV as well as window orientations in the high insulation scenario, in Helsinki: (**a**) North; (**b**) South; (**c**) East; (**d**) West.

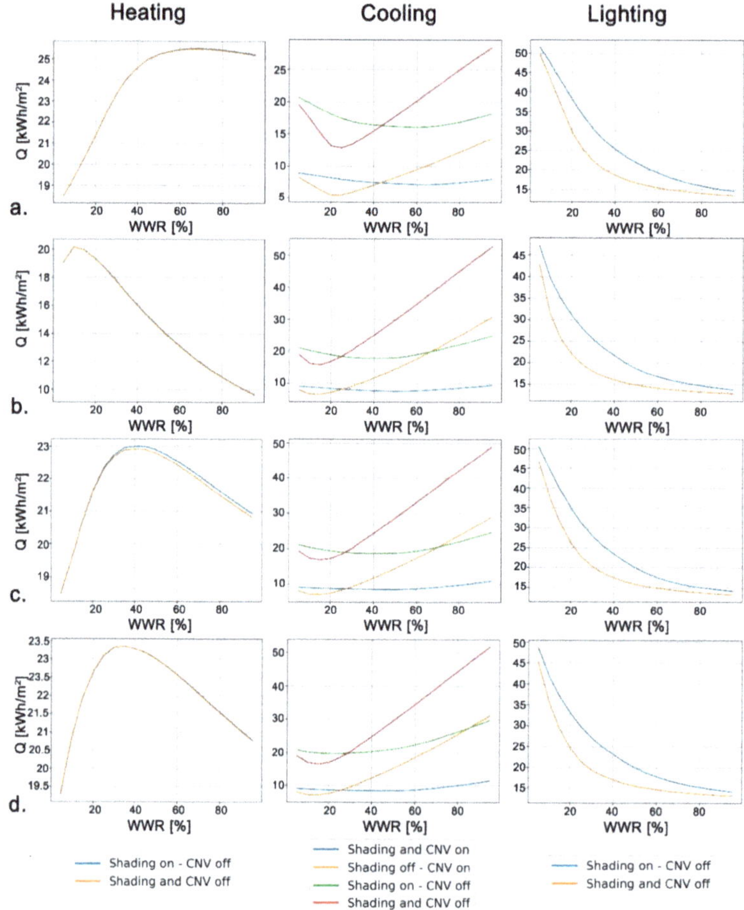

Figure 8. Heating, Cooling, and lighting annual energy needs as a function of WWR for different setting of shading and CNV as well as window orientations in the high insulation scenario, in Turin: (**a**) North; (**b**) South; (**c**) East; (**d**) West.

4. Discussion

4.1. Monthly Energy Needs

The monthly energy need distribution as a dependent variable of WWR for each considered window shading/opening configuration allows for a more detailed assessment of the simulation results. These data are shown in Figures 9 and 10 for Helsinki and Turin, respectively, considering the heating and cooling needs—using a variation range of WWR by intervals of 5%. As the graphs show, the general trends follow the expected distributions: high heating values of energy needs during the winter season and low or equal to zero values during the summer, while an opposite trend is highlighted for cooling energy needs. However, it is possible to state that varying WWR has an effect on energy needs as described in the general comments of Figures 7 and 8. Focusing on the high insulated scenario, in the heating season, the distribution of monthly energy needs shows that for colder months: low WWR are suitable, while during other winter months high WWR may perform better. This occurs mainly with the South-facing window and not with the North-facing one due to the almost null potential of solar gains—see Figure 9. For example, in the Helsinki South-facing case, the reversal between high and low WWR as an optimal configuration, occurs between January and February, and between October and November, in the high-insulated scenario. Differently, for the cooling season, when shading and CNV are not activated, high WWRs show the worst behaviour in all months—see Figure 10—while this trend is counterbalanced when CNV is activated.

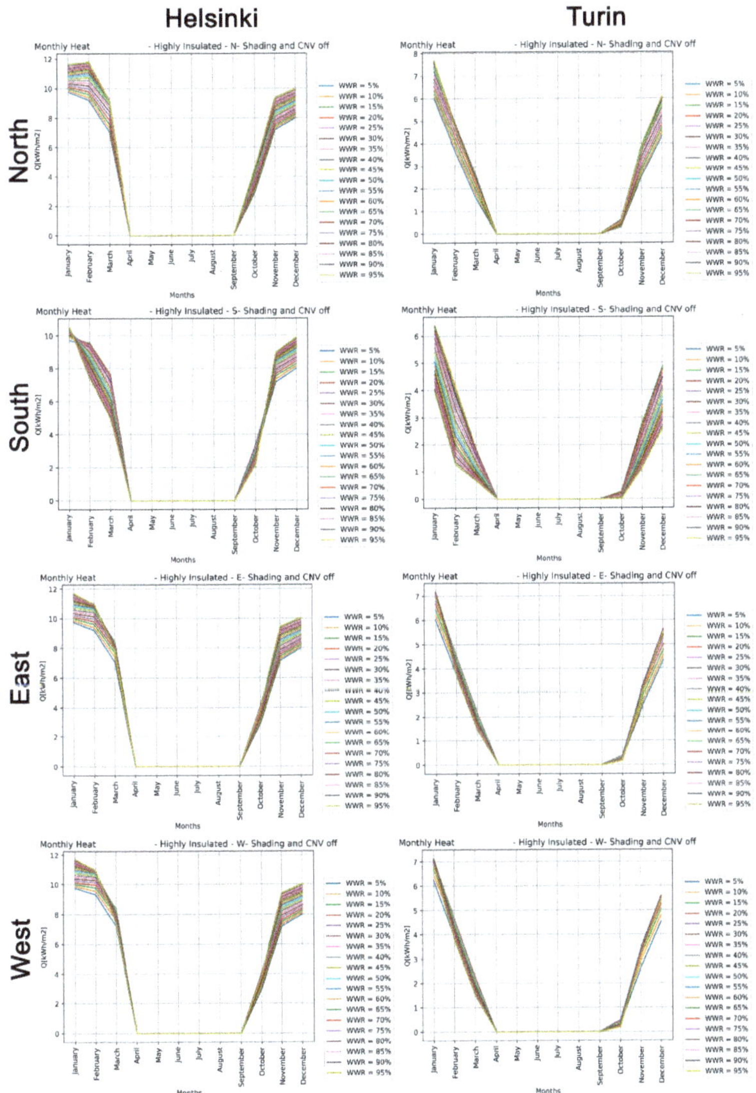

Figure 9. Monthly distribution of the heating energy needs with Shading and CNV "Off"—high insulated scenario.

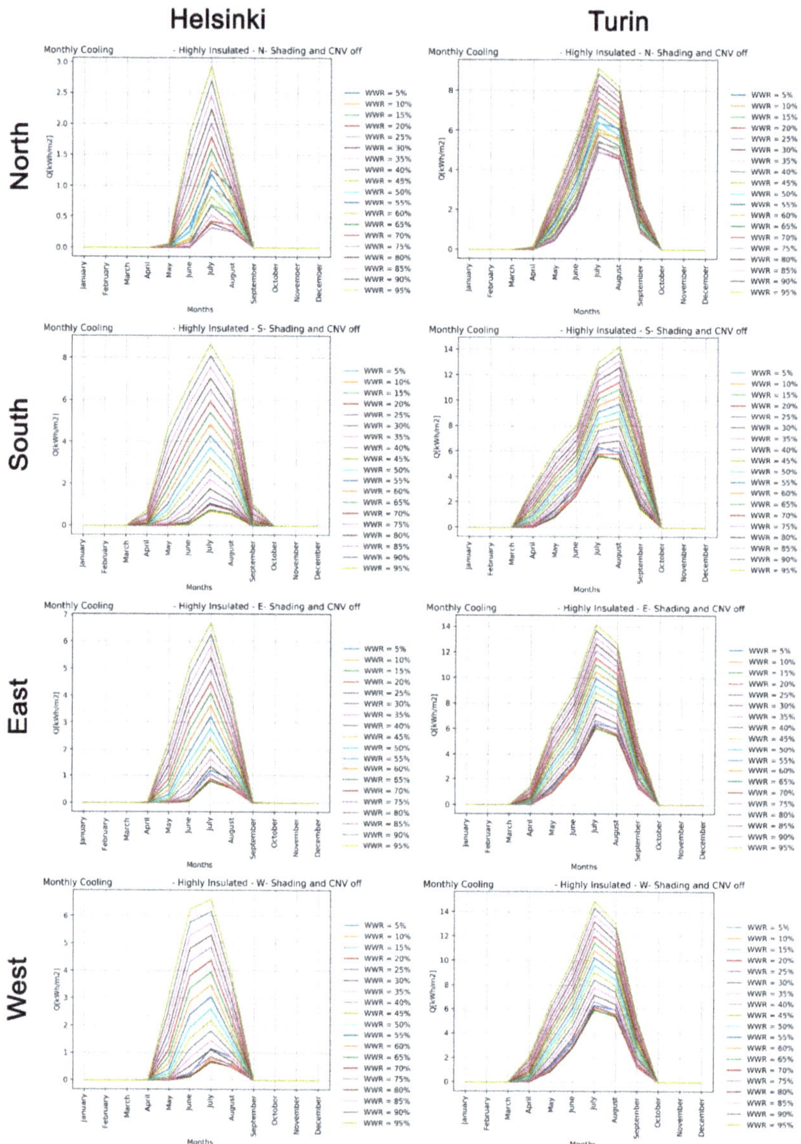

Figure 10. Monthly distribution of the cooling energy needs with Shading and CNV "Off"—high insulated scenario.

When both shading and CNV systems are activated—see Figure 11—the monthly cooling energy needs decrease considerably due to the heat dissipation effect of ventilative cooling and the heat gain prevention of shading. Moreover, this effect is not only apparent in terms of the intensity of the cooling need, but also in the number of months where the cooling system has to be activated. In fact, as is shown in Figure 11, it is possible to underline that these passive cooling systems may almost nullify the cooling needs in the Helsinki case, passing from a peak of about 8.5 kWh/m^2 (WWR 95%) for the South façade, to a peak of 0.2 kWh/m^2. Furthermore, the number of months interested by cooling needs, drastically reduces. Similarly, in the Turin case, the effect of shading and ventilative cooling more than half the cooling energy needs in all façade orientations. For the south façade case, in particular, cooling needs decrease from about 14.2 kWh/m^2 to about 6 kWh/m^2 (WWR 95%). Considering the number of cooling months, a reduction from the April-September period in the case without passive solutions, and May/June-August period with these counteractions (south-façade) is evident.

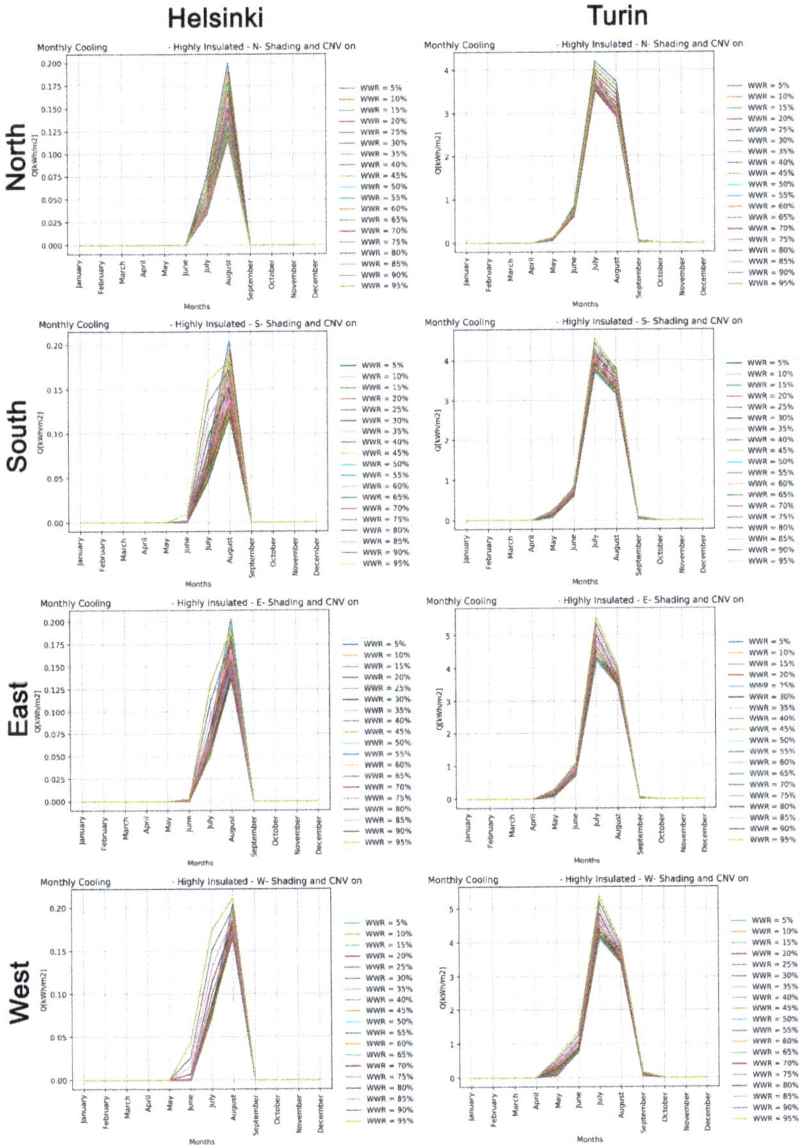

Figure 11. Monthly distribution of the cooling energy needs with Shading and CNV "On"—high insulated scenario.

This could be explained by the greenhouse effect happening during daylight hours, especially for the highly insulated scenario. The differences between Helsinki and Turin, in term of cooling intensity and number of months when cooling is needed, are related to local climate characteristics. In a temperate climate such as Turin's, overheating may occur also in winter months, particularly in high insulated buildings [38]. CNV and shading do not affect heating needs due to the adopted activation thresholds, except for a very little impact in some spring and fall months.

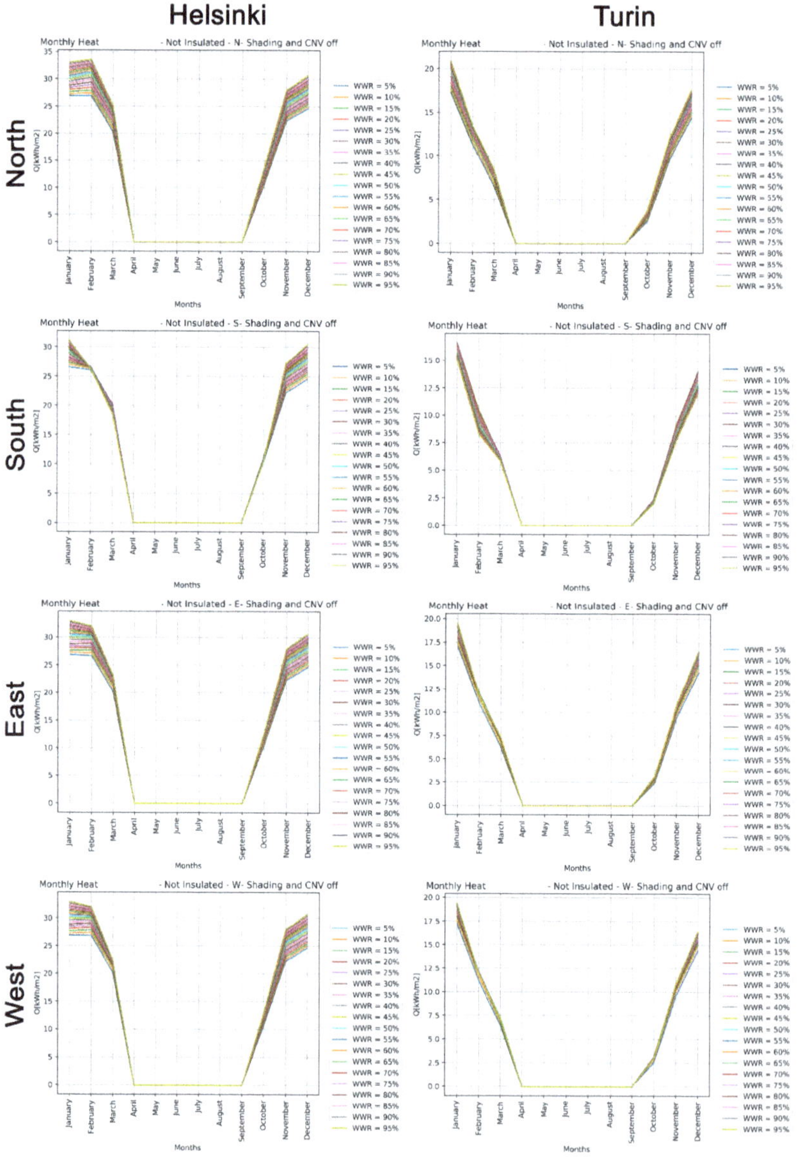

Figure 12. Monthly distribution of the heating energy needs with Shading and CNV "Off"—not-insulated scenario.

When the non-insulated scenario is considered, the general heating and cooling energy trends are comparable to the ones of the high-insulated building, but with a difference in absolute energy intensity values—see Figures 12 and 13. Nevertheless, small changes may be found for cooling energy needs in the case of CNV and shading—see Figure 14—as demonstrated by the results of the annual analysis—see Figures 7 and 8.

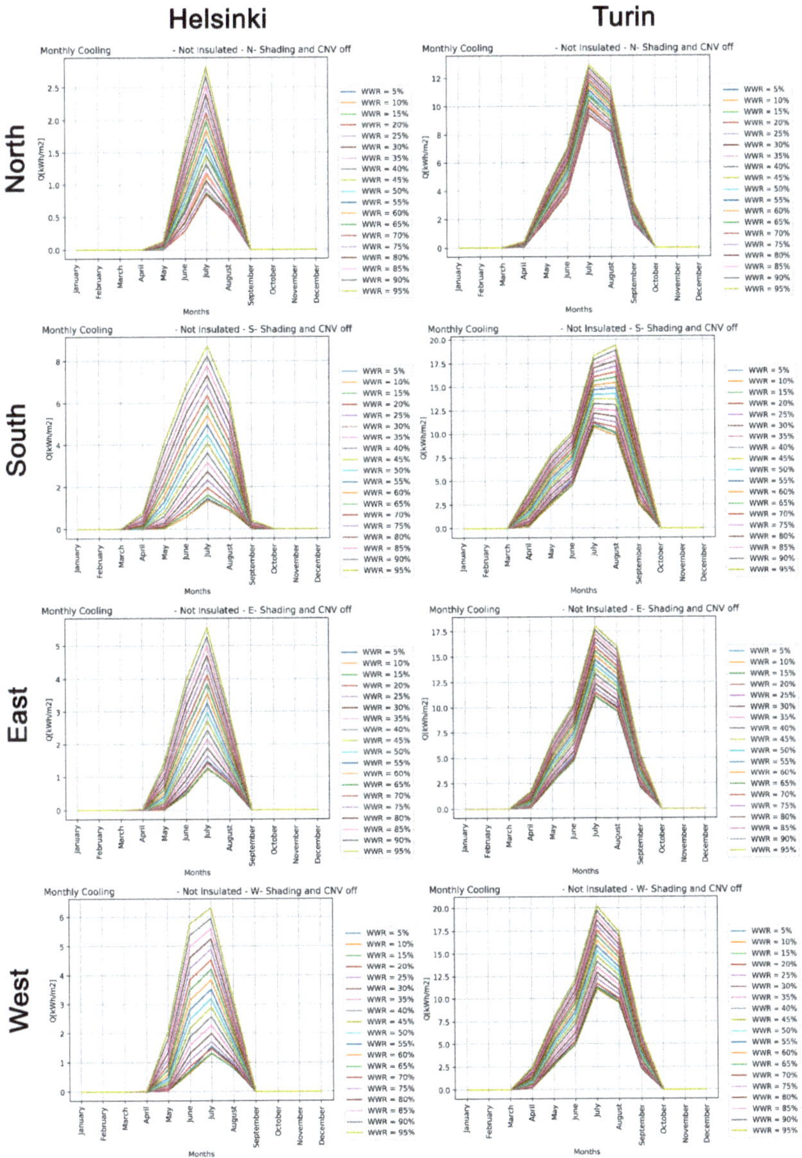

Figure 13. Monthly distribution of the cooling energy needs with Shading and CNV "Off"—not-insulated scenario.

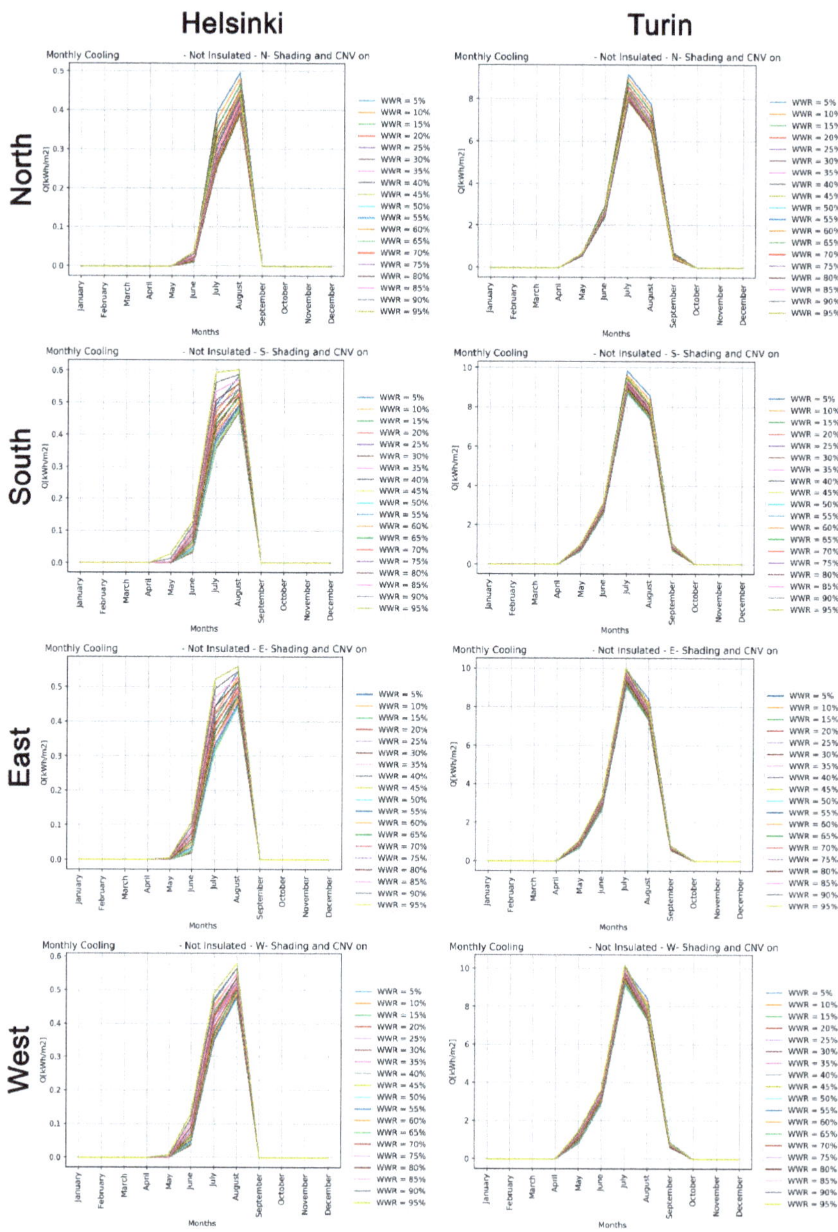

Figure 14. Monthly distribution of the cooling energy needs with Shading and CNV "On"—not-insulated scenario.

4.2. Sensibility Analysis by Changing the Occupancy Value

In this section, the methodology and results of a sensibility analysis carried out with the changing occupancy rate dynamically in every simulation are described in order to evaluate the impact of the random presence of people on energy needs. Based on the daily schedule of an office building, or general occupancy, each simulation was performed, inputting a random value of people density derived from a Gaussian distribution, $G(\mu,\sigma)$, and assuming mean and variance values as described in the methodological Section 2. Hence, combining all variables as described in Table 1, a total of 48 cases for each location, resulting in 16,128 simulations were carried out.

Some results, expressed as the total energy needs as a function of WWR, are shown in Figures 17 and 18, in relation to the following configurations.

Heating season:

Insulation scenario: Highly Insulated
Exposure: South; North
Shading and CNV Setup: both "Off"

Cooling season:

Insulation scenario: Highly Insulated
Exposure: South
Shading and CNV Setup: both "Off; both "On"

For each value of WWR, 16 different values of energy need intensity [kWh/m^2] were calculated by assuming relevant random occupancy variations. As expected, an increase in the number of people led to a decrease of heating energy need and to an increase of cooling need. Nevertheless, the decrease of heating energy need is not as sharp as the increase of the cooling need, partially due to the clothing schedule used in the software.

The energy need for lighting does not have any random variation but is related to the illuminance requirement set for an office and relevant schedule.

4.2.1. Heating Energy Need

Figure 15 reports the heating energy needs for Helsinki and Turin for the 2 defined configurations (North and South facing window). In both cases, the heating energy need is highly influenced by the occupancy variation, depending on the relevant internal gain variation; this is more apparent for Helsinki due to the lower ambient temperature. Nevertheless, after an initial negative effect, when an increase of the average façade U-value is not sufficiently balanced by an increase of solar gains, the increment of WWR allows for reducing the heating demand. This is true for Turin, due to temperate climate conditions, while it is less apparent in Helsinki, in line with other studies [26]. When the window is facing north, the smallest values of heating energy needs correspond to the lowest WWR in both Helsinki and Turin, due to the limited amount of solar gains reaching north-facing façades in winter.

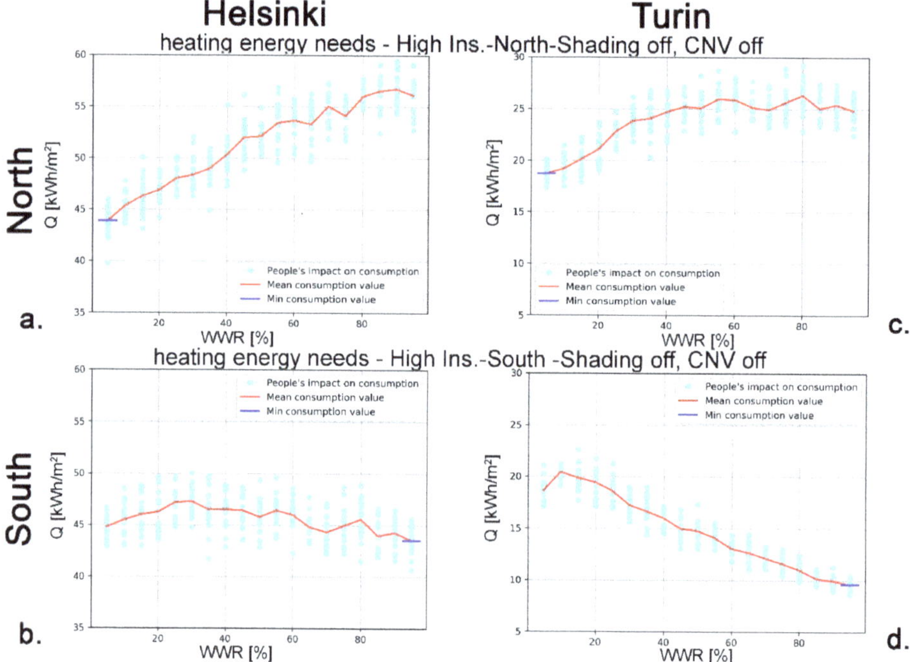

Figure 15. Heating energy needs for the highly insulated building, CNV and Shading off—for Helsinki (**a**) North-facing case; (**b**) South-facing case; and for Turin (**c**) North-facing case; (**d**) South-facing case.

4.2.2. Cooling Energy Need

Cooling energy needs increase with WWR in both locations when CNV and shading are not activated—see Figure 16. Their trend is similar in both locations, while their absolute intensity values are remarkably different due to local climate conditions. When switching "On" both shading and CNV, the energy need decreases considerably until about 55% of WWR in both locations. After this value, the cooling need start to grow again because the cooling effect of CNV and shading is not sufficient to counterbalance the heat due to solar and internal gains. Roughly speaking, the average trends are similar to the ones shown in Figures 9 and 10, even if the effect of random occupancy on the internal heat gain may slightly alter the results. Nevertheless, the random presence of people does not vary so much the cooling energy needs for each WWR in comparison to its effect in the heating season (CNV and shading set to Off)—the cooling variance is, in fact, smaller than in the heating case.

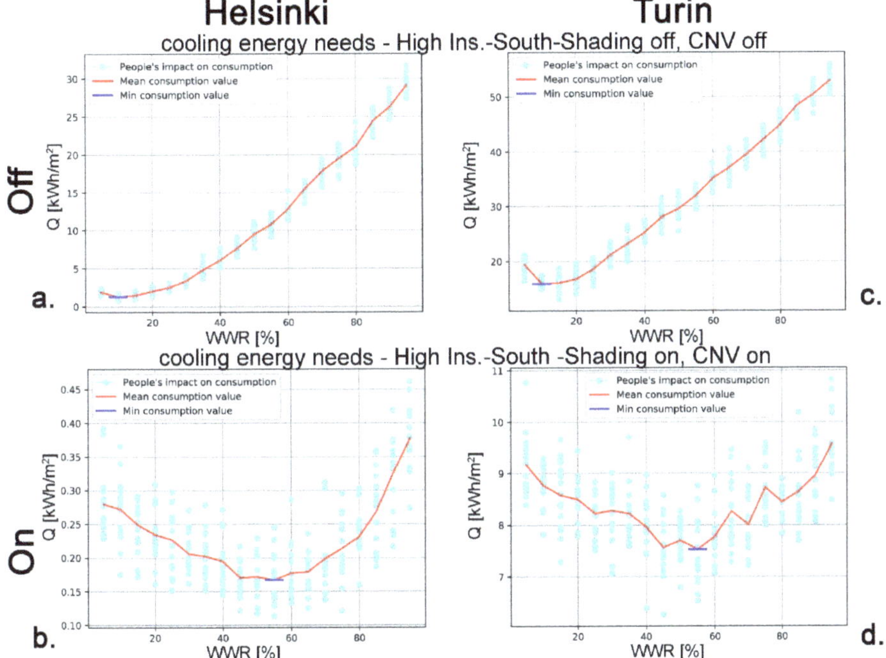

Figure 16. Cooling energy needs for the highly insulated building, South-facing case—for Helsinki (**a**) CNV and shading off; (**b**) CNV and shading on; and for Turin (**c**) CNV and shading off; (**d**) CNV and shading on.

4.2.3. Total Energy Needs

If CNV and shading are off, the optimum WWR for a South-facing window, corresponding to the lowest energy need, is reached between 35–40% for Helsinki and at about 30% in Turin, considering the random occupancy effect—see Figure 17. The ventilative cooling effect in reducing the cooling needs as well as the shading effect in preventing the solar gains are apparent both in Helsinki and Turin, with differences related to the local impact of cooling loads, higher in Turin than in Helsinki. When CNV and shading are activated, the highest WWR corresponds to the lowest energy need in both cases due to the possibility of balancing summer overheating without compromising the positive effect of winter solar gain. In fact, thanks to the local climate conditions of the considered locations, summer outdoor air temperatures are sufficiently lower than both the indoor and the comfort threshold temperatures.

If CNV and shading are off, the optimum WWR for a North-facing window is reached between 35–40% for both Helsinki and Turin—see Figure 18. Even in this case, the positive effect of shading and CNV in reducing the cooling demand is apparent. In particular, for Helsinki the optimal WWR is reached for values around 85%, while for Turin, values around 95% are suggested. This difference is due to the yearly balance between heating and cooling energy needs according to local climate conditions.

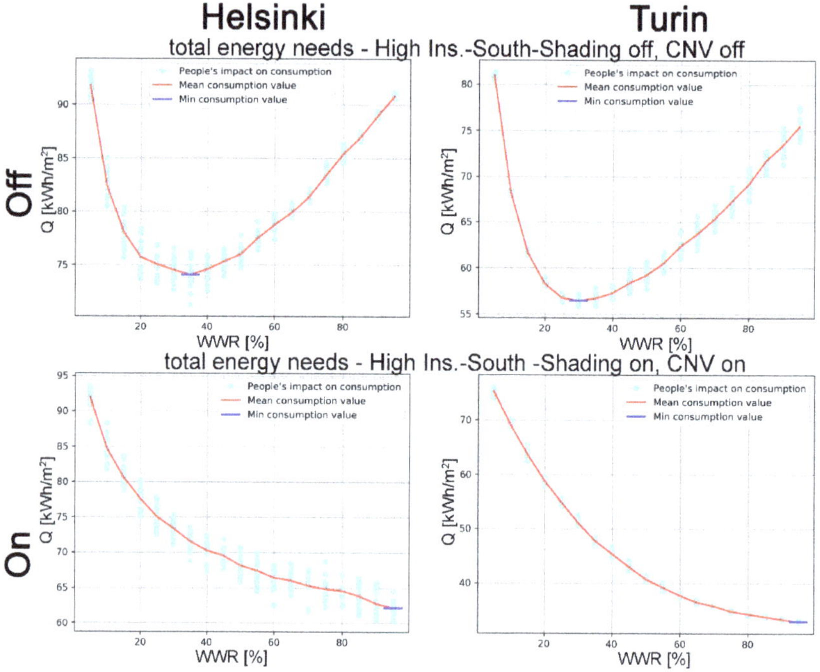

Figure 17. Total energy needs for a highly insulated building, South, Shading and CNV "On-Off".

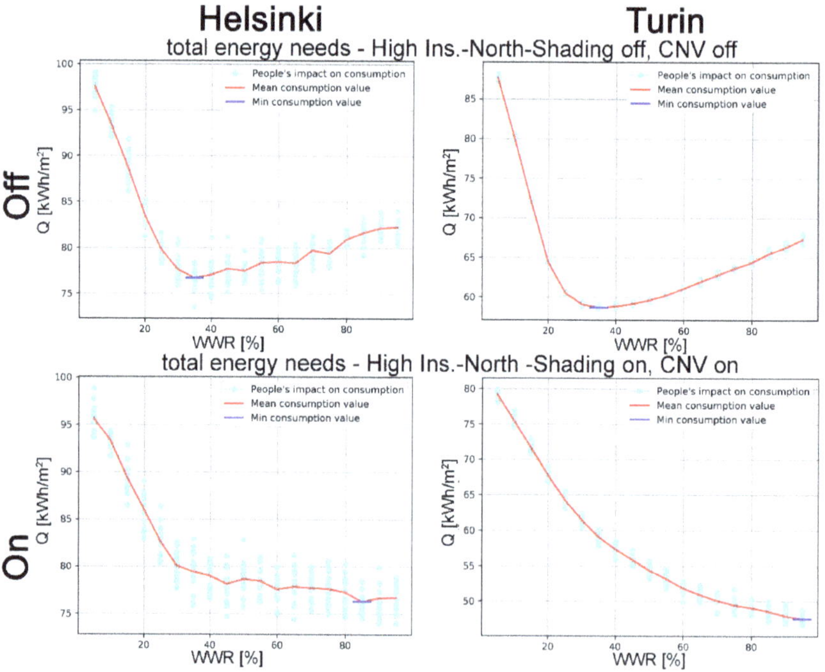

Figure 18. Total energy needs for a highly insulated building, North, Shading and CNV "On-Off".

4.3. Regression

Starting from the number of points derived from the simulations, a dataset was built in order to fit a model that could predict the energy need based on WWR. An assumption is made that those points are the independent variables (predictors), with values varying at every 5% interval, and as a dependent variable the energy need intensity Q (kWh/m^2). The polynomial curve fitting method implemented in the Numpy Python library was used as regression technique. This technique approximates the process of constructing a curve, or a mathematical function that has the best fit to a series of data points. This technique works in both the case in which data on the y axis has shape (1,1) and the case where shape is (N,1). Here, N = 16. A number of alternative curves, with a degree of the polynomial fitting ranging from 1 to 6, were then elaborated and the relevant RMSE (root mean square error) calculated.

The RMSE values are useful to select the best fitting curve, i.e., choosing a polynomial degree avoiding both underfitting and overfitting problems. If the degree of the fitting curve is too low—underfitting—then the fitting curve is missing important features, while if the degree is too high—overfitting—then the fitting curve is also modelling noise.

The formula used to calculate the RMSE is the following, where predicted and target are N-dimensional vectors.

$$\text{RMSE} = (\text{mean}((\text{predicted-target})^2)^{0.5} \tag{1}$$

The RMSE was calculated by comparing the mean of the Gaussian distribution, from which the values used for getting the random occupancy points was derived (test set), against the curve derived from varying the WWR while keeping the occupation density constant—no added random noise—(training set). The obtained RMSE can be considered as an evaluation of the similarity of the regressed curve over the training set, because the energy need for these WWR values was known since they had been simulated.

In addition, for evaluating the accuracy of the regression model, it was also possible to simulate, for a specific setup, the heating and cooling energy needs for different values of WWR that were not considered in the training domain. This analysis was performed for the Helsinki case, even if the same method may be applied to different locations.

4.3.1. Regression over the Train Set

Regression analyses were not performed on the lighting energy need because random occupancy variations do not influence this specific value, being dependent on a fixed illuminance threshold and on the percentage of natural light passing through the window, but independent of the intensity of people present. Differently, cooling and heating energy needs are dependent on the random occupancy variation due to people internal gain production. Nevertheless, the analyses on the total energy needs are based on cooling, heating and lighting.

For the configurations shown in Figure 19, the polynomial degree that fits best the extracted points is 4 for both heating and cooling. On the contrary, for the total energy, the polynomial degree that fits best the extracted points is the highest—8—because of the behaviour of the dataset distribution.

As expected, the RMSE values are very small. In fact, the points used to fit the model and to perform the regression can be considered as noised values of the curves that we use to calculate the error. A test on an independent database is hence needed.

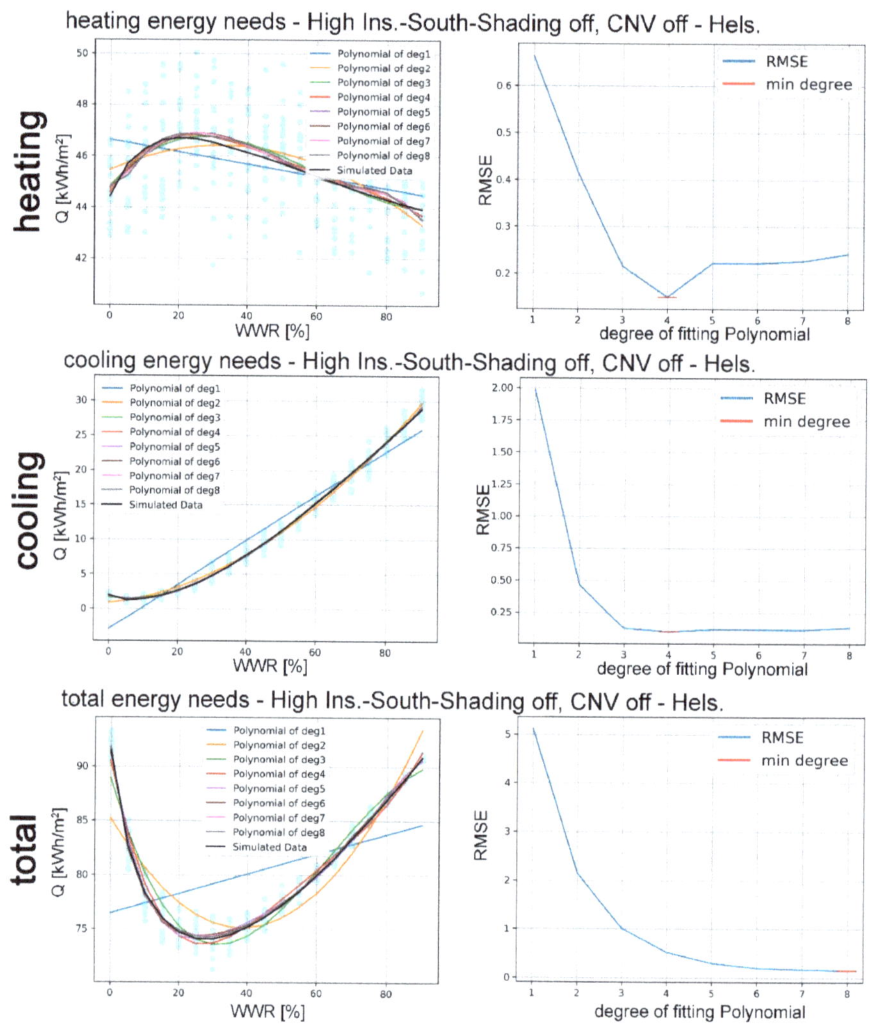

Figure 19. Figure 17—Heating, Cooling, Total regressed energy needs, with the corresponding RMSE for each degree (training).

4.3.2. Regression over the Test Set

In this second test analysis, a prediction of energy needs for heating and cooling was elaborated, based on unknown values of WWR.

Starting from the model derived from the train dataset – WWR = [1, 5, 10, 15, … , 95]—a regression analysis on new test values—WWR = [2.5, 7.5, 12.5, …]—was carried out. As can be seen in Figure 20—with CNV and shading off, and in Figure 21 with CNV and shading on, energy needs are well predicted also when using other values of WWR. Comparing these values to the ones those derived from simulating the new values of WWR by EnergyPlus, the RMSE was calculated. As expected, this RMSE is a little bit higher than the RMSE of the training set, even if is still very low. The degree of the polynomial that better fits the points is again the fourth for the heating and the cooling cases.

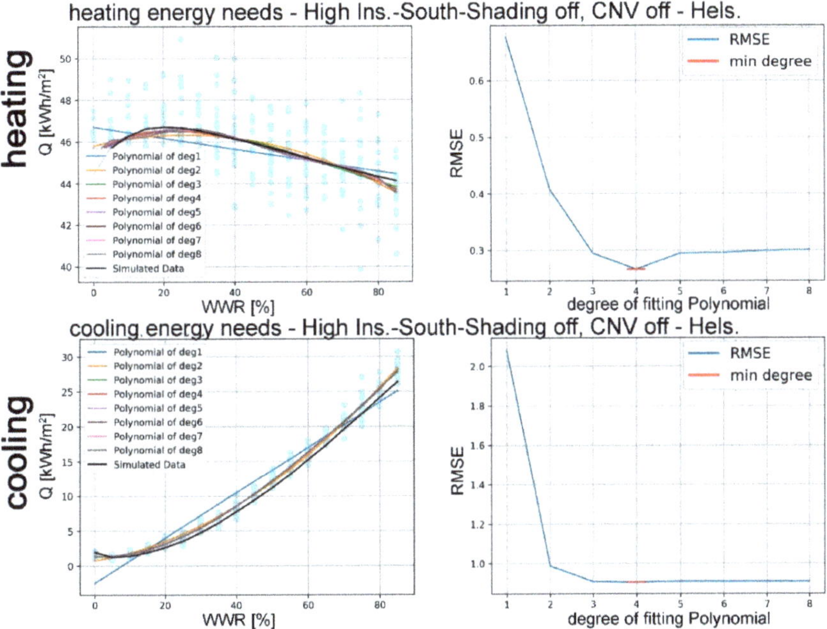

Figure 20. Heating and Cooling regressed energy needs, with the corresponding RMSE for each degree (testing case)—CNV and shading off setup.

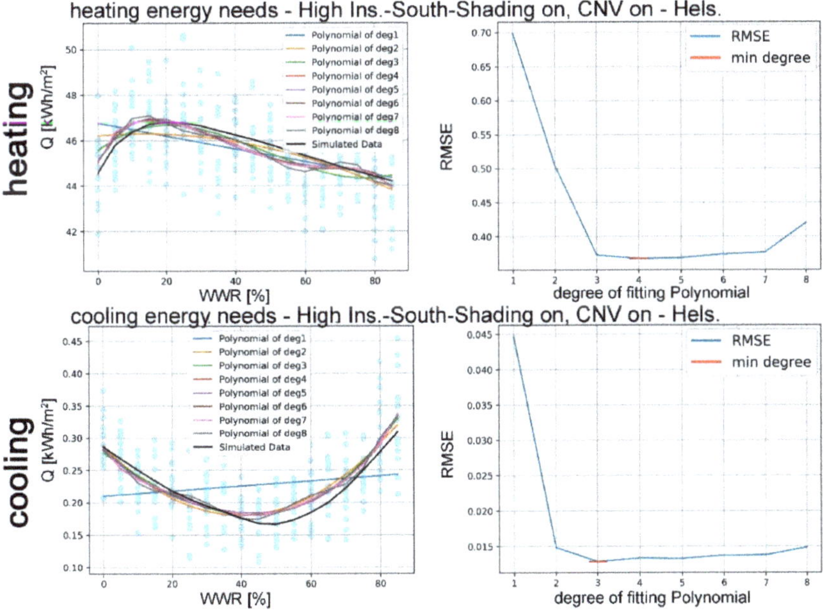

Figure 21. Heating and Cooling regressed energy needs, with the corresponding RMSE for each degree (testing case)—CNV and shading on setup.

In Tables 5 and 6, the values of RMSE for the training and testing of cooling and heating energy are shown.

Table 5. RMSE Cooling for Training and Testing.

	RMSE Cooling train	**RMSE Cooling test**	**Deg Train**
High Ins N Shading and CNV on	0.0076	0.1282	5
High Ins N Shading and CNV off	0.1004	0.9049	4

Table 6. RMSE Heating for Training and Testing.

	RMSE Heating Train	**RMSE Heating test**	**Deg Train**
High Ins S Shading and CNV on	0.2417	0.3676	6
High Ins S Shading and CNV off	0.1500	0.2663	4

5. Conclusions

The analyses presented in this paper help to characterize, through hourly-based dynamic simulations, the influence of the window-to-wall ratio (WWR) on the energy need for space heating and cooling, and the lighting of an office building in two reference locations representing a cold and a moderate climate zone of Europe. Various envelope and window characteristics were considered as independent variables, namely: insulation level, orientation, shading, controlled natural ventilation.

Results of simulations at constant occupation rate show that an optimal WWR value, balancing the three above-mentioned energy uses in terms of the least energy annual need, can be found around 30% for both locations.

A second type of study dealt with a regression analysis of data resulting from simulations carried out by an algorithm developed for the purpose of allowing a changing occupation rate based on random behaviour. This analysis aimed to simulate the reference case conditions in the way closest to the actual dynamic context. In addition, several regression curves and the relevant RMSE were yielded and compared in order to find the correlation factor which could best fit the analysed data sets.

In general, the analyses carried out have a methodological value in representing an innovative approach to define the optimal configurations of building envelopes and window characteristics with respect to the minimization of annual energy needs. Furthermore, this study's purpose is consistent with the minimum requirements of a passive house and includes the potential effects of random variation in occupancy. This approach allows for supporting design choices from the preliminary phase while considering perturbation phenomena that may occur in real situations.

Author Contributions: Article conceptualization, G.C.; methodology, G.C., A.A.; software, L.B., M.F., E.P., E.S.; investigation, G.C., M.F., E.P., E.S.; writing—original draft preparation, G.C., M.F., E.P., E.S.; writing—review and editing, G.C., M.G.; supervision, G.C., with A.A., M.G.; funding acquisition, G.C.

Funding: The research was co-funded by the 59_ATEN_RSG16CHG.

Acknowledgments: The proposed approach was tested during the Course ICT in Building Design, Master Degree in ICT for Smart Society, Politecnico di Torino, Italy, A.Y. 2018-19, with the support of the LASTIN laboratory, Microclimate section.

Conflicts of Interest: The authors declare no conflict of interest. The funders had no role in the design of the study; in the collection, analyses, or interpretation of data; in the writing of the manuscript, or in the decision to publish the results.

Nomenclature

WWR	Windows-to-Wall Ratio
VC	Ventilative Cooling
CNV	Control Natural Ventilation
RMSE	Root Mean Square Error
DB	Design Builder software

References

1. Orme, M. Estimates of the energy impact of ventilation and associated financial expenditures. *Energy Build.* **2011**, *33*, 199–205. [CrossRef]
2. Cuce, P.M.; Riffat, S. A state of the art review of evaporative cooling systems for building applications. *Renew. Sustain. Energy Rev.* **2016**, *54*, 1240–1249. [CrossRef]
3. Logue, J.M.; Sherman, M.H.; Walker, I.S.; Singer, B.C. Energy impacts of envelope tightening and mechanical ventilation for the U.S. residential sector. *Energy Build.* **2013**, *65*, 281–291. [CrossRef]
4. European Commission. *Communication from the Commission to the European Parliament, the Council, the European Economic and Social Committee, the Committee of the Regions and the European Investment Bank, Clean Energy for All Europeans*; COM(2016) 860 Final; European Commission: Brussels, Belgium, 2016.
5. Santamouris, M. (Ed.) *Advances in Passive Cooling*; Earthscan: London, UK, 2007; ISBN 978-1-84407-263-7.
6. Santamouris, M.; Asimakopolous, D. (Eds.) *Passive Cooling of Buildings*; James & James: London, UK, 1996; ISBN 1-873936-47-8.
7. IEA. *The Future of Cooling. Opportunities for Energy Efficient Air Conditioning*; International Energy Agency: Paris, France, 2018.
8. Kolokotroni, M.; Heiselberg, P. (Eds.) *Ventilative Cooling State of the Art*; International Energy Agency—Energy in Buildings and Communities Programme, Aalborg University: Aalborg, Denmark, 2015; ISBN 87-91606-25-X.
9. Grosso, M.; Acquaviva, A.; Chiesa, G.; da Fonseca, H.; Bibak Sareshkeh, S.S.; Pardilla, M.J. Ventilative cooling effectiveness in office buildings: A parametrical simulation. In Proceedings of the 39th AIVC—7th TightVent & 5th Venticool Conference, Antibes Juan-Les-Pins, France, 18–19 September 2018; AIVC-INIVE: Brussels, Belgium, 2019; pp. 780–788.
10. Chiesa, G.; Grosso, M.; Pearlmutter, D.; Ray, S. Editorial. Advances in adaptive comfort modelling and passive/hybrid cooling of buildings. *Energy Build.* **2017**, *148*, 211–217. [CrossRef]
11. Santamouris, M. Cooling the buildings—Past, present and future. *Energy Build.* **2016**, *28*, 617–638. [CrossRef]
12. Goia, F.; Haase, M.; Perino, M. Optimizing the configuration of a façade module for office buildings by means of integrated thermal and lighting simulations in a total energy perspective. *Appl. Energy* **2013**, *108*, 515–527. [CrossRef]
13. Arumi, F. Day lighting as a factor in optimizing the energy performance of buildings. *Energy Build.* **1977**, *1*, 175–182. [CrossRef]
14. Johnson, R.; Sullivan, R.; Selkowitz, S.; Nozaki, S.; Conner, C.; Arasteh, D. Glazing energy performance and design optimization with daylighting. *Energy Build.* **1984**, *6*, 305–317. [CrossRef]
15. Baker, N.V.; Steemers, K. LT Method 3.0—A strategic energy-design tool for Southern Europe. *Energy Build.* **1996**, *23*, 251–256. [CrossRef]
16. Su, X.; Zhang, X. Environmental performance optimization of window-wall ratio for different window type in hot summer and cold winter zone in China based on life cycle assessment. *Energy Build.* **2010**, *42*, 198–202. [CrossRef]
17. Ma, P.; Wang, L.-S.; Guo, N. Maximum window-to-wall ratio of a thermally autonomous building as a function of envelope U-value and ambient temperature amplitude. *Appl. Energy* **2015**, *146*, 84–91. [CrossRef]
18. Lee, J.W.; Jung, H.J.; Park, J.Y.; Lee, J.B.; Yoon, Y. Optimization of building window system in Asian regions by analysing solar heat gain and daylighting elements. *Renew. Energy* **2013**, *50*, 522–531. [CrossRef]
19. Kheir, F. The relation of orientation and dimensional specifications of window with building energy consumption in four different climates of Köppen classification. *Researcher* **2013**, *5*, 107–115.
20. Goia, F. Search for the optimal window-to-wall ration in office buildings in different European climates and the implications on total energy saving potential. *Sol. Energy* **2016**, *132*, 467–492. [CrossRef]
21. Košir, M.; Gostiša, T.; Kristl, Z. Influence of architectural building envelope characteristics on energy performance in Central European climatic conditions. *J. Build. Eng.* **2017**, *15*, 278–288. [CrossRef]
22. Echenagucia, T.M.; Capozzoli, A.; Cascone, Y.; Sassone, M. The early design stage of a building envelope: Multi-objective search through heating, cooling and lighting energy performance analysis. *Appl. Energy* **2015**, *154*, 577–591. [CrossRef]
23. Consumption of Buildings. A parametric analysis in Italian climate conditions. *J. Build. Eng.* **2017**, *13*, 169–183. [CrossRef]

24. Alghoul, S.K.; Rijabo, H.G.; Mashena, M.E. Energy consumption in buildings: A correlation for the influence of window to wall ratio and window orientation in Tripoli, Libya. *J. Build. Eng.* **2017**, *11*, 82–86. [CrossRef]
25. Wen, L.; Hiyama, K.; Koganei, M. A method for creating maps of recommended window-to-wall ratios to assign appropriate default values in design performance modeling: A case study of a typical office building in Japan. *Energy Build.* **2017**, *145*, 304–317. [CrossRef]
26. Feng, G.; Chi, D.; Xu, X.; Dou, B.; Sun, Y.; Fu, Y. Study on the Influence of Window-wall Ratio on the Energy Consumption of Nearly Zero Energy Buildings. *Procedia Eng.* **2017**, *205*, 730–737. [CrossRef]
27. Sun, Y.; Shanks, K.; Baig, H.; Zhang, W.; Hao, X.; Li, Y.; He, B.; Wilson, R.; Liu, H.; Sundaram, S.; et al. Integrated CdTe PV gazing into windows: Energy and daylight performance for different window-to-wall ratio. *Energy Procedia* **2019**, *158*, 3014–3019. [CrossRef]
28. Xue, Q.; Li, Q.; Xie, J.; Zhao, M.; Liu, J. Optimization of window-to-wall ratio with sunshades in China low latitude region considering daylighting and energy saving requirements. *Appl. Energy* **2019**, *233–234*, 62–70. [CrossRef]
29. Chiesa, G.; Grosso, M. An Environmental Technological Approach to Architectural Programming for School Facilities. In *Mediterranean Green Buildings & Renewable Energy*; Sayigh, A., Ed.; Spinger: Cham, Switzerland, 2017; pp. 701–716. [CrossRef]
30. Passive House Institute. Available online: https://passivehouse.com/index.html (accessed on 10 May 2019).
31. Chiesa, G.; Grosso, M.; Acquaviva, A.; Makhlouf, B.; Tumiatti, A. Insulation, building mass and airflows provisional and multivariable analysis. *Sustain. Mediterr. Constr.—SMC* **2018**, *8*, 36–40.
32. The British Council for Offices. *Occupier Density Study 2013*; BCO: London, UK, 2013.
33. Grosso, M. (Ed.) *Il raffrescamento passivo degli edifici*, 2nd ed.; Maggioli: Sant'Arcangelo di Romagna, Italy, 2008; p. 313, ISBN 978-88-387-3963-3.
34. Olgyay, A.; Olgyay, V. *Solar Control and Shading Devices*; Princeton University Press: Princeton, NJ, USA, 1957.
35. U.S. Department of Energy. *EnergyPlus™ Version 8.9.0 Documentation. Engineering Reference*; U.S. Department of Energy: Washington, DC, USA, 2018.
36. Watson, D.; Labs, K. *Climatic Design. Energy-Efficient Building Principles and Practices*; McGraw-Hill: New York, NY, USA, 1983; ISBN 0-07-068478-2.
37. Mazria, E. *The Passive Solar Energy Book*; Rodale: Emmaus, PA, USA, 1979; ISBN 0-87857-237-6.
38. Heiselberg, P. (Ed.) *Ventilative Cooling Design Guide*; IEA EBC Annex 62; Aalborg University Press: Aalborg, Denmark, 2018; ISBN 87-91606-38-1.

Article

Multisensor IoT Platform for Optimising IAQ Levels in Buildings through a Smart Ventilation System

Giacomo Chiesa [1,*], Silvia Cesari [2], Miguel Garcia [2], Mohammad Issa [2] and Shuyang Li [2]

[1] Department of Architecture and Design, Politecnico di Torino, 10125 Turin, Italy
[2] ICT for Smart Society, Department of Electronics and Telecommunications, Politecnico di Torino, 10129 Turin, Italy; silvia.cesari@studenti.polito.it (S.C.); miguel.garcia@studenti.polito.it (M.G.); mohammad.issa@studenti.polito.it (M.I.); shuyang.li@studenti.polito.it (S.L.)
* Correspondence: giacomo.chiesa@polito.it; Tel.: +39-011-090-4376

Received: 12 July 2019; Accepted: 15 October 2019; Published: 18 October 2019

Abstract: Indoor Air Quality (IAQ) issues have a direct impact on the health and comfort of building occupants. In this paper, an experimental low-cost system has been developed to address IAQ issues by using a distributed internet of things platform to control and monitor the indoor environment in building spaces while adopting a data-driven approach. The system is based on several real-time sensor data to model the indoor air quality and accurately control the ventilation system through algorithms to maintain a comfortable level of IAQ by balancing indoor and outdoor pollutant concentrations using the Indoor Air Quality Index approach. This paper describes hardware and software details of the system as well as the algorithms, models, and control strategies of the proposed solution which can be integrated in detached ventilation systems. Furthermore, a mobile app has been developed to inform, in real time, different-expertise-user profiles showing indoor and outdoor IAQ conditions. The system is implemented in a small prototype box and early-validated with different test cases considering various pollutant concentrations, reaching a Technology Readiness Level (TRL) of 3–4.

Keywords: indoor air quality index; smart ventilation; IoT; microservice; mobile app; fan-assisted ventilation; indoor air quality

1. Introduction

Sick Building Syndrome (SBS) affects people who are subject to extended exposure to chemical, biological, and/or physical agents found in buildings with a low Indoor Air Quality (IAQ) level which is generally related to bad ventilation. Moreover, the World Health Organization (WHO) reported that IAQ is generally from 2 to 5 times worse than outdoor air quality thus increasing the risk of both short term effects, e.g., eye irritation or a decrease in productivity, and long term effects, e.g., increased risk of headaches, damage to central nervous system, or even cancer [1]. Furthermore, current life styles cause an increase in the number of hours spent indoors, taking up about 90% of our time [2,3]. Bad IAQ levels are related to several factors, including dampness and mould—see also [4]—and other pollutant concentrations in enclosed spaces. In Iceland, Switzerland and Norway the number of people affected by dampness was estimated to be about 80.4 million [3], while the WHO confirms that about 3.8 million people died annually due to household air pollution at a world level [5]. The study of low cost technologies to monitor IAQ levels and prevent, by actuating actions, the persistence of critical conditions is an essential part of this research. Especially in highly insulated buildings, such as the ones that use the passivehaus approach, a mechanical/hybrid ventilation system is needed [6], while a smart activation of the same is suggested to prevent excessive consumption. Furthermore, when this approach, principally adapted for cold locations, is extended to temperate climatic conditions, solutions to guarantee low-cost ventilation are needed to reduce overheating [7]. The solution proposed in this paper is adapted to face this challenge and can be integrated with detached and traditional mechanical systems or used alone in a hybrid, natural prevalent ventilation solution, with special regard to residential properties and small offices or to public buildings (e.g., school rooms).

The sources of indoor pollution include combustion of solid fuels in indoor environments, especially in developing countries; tobacco smoke; outdoor air pollutants; construction materials and furnishing sources [8]; bad maintenance of ventilative and air conditioning systems; but also cleaning products, cosmetics, and equipment (e.g., printers)—see also [9]. The main indoor pollutant substances are Carbon dioxide (CO_2), Carbon monoxide (CO), Nitrogen oxides (NOx), Ozone (O_3), Pentachlorophenol (PCP), Polycyclic Aromatic Hydrocarbon (PAH), Tobacco smoke, and Volatile Organic Compounds (VOCs), but also PM values inherited from outdoor air or from indoor fuel combustion, e.g., [10], polycyclic aromatic hydrocarbons; radon and other similar pollutants are considered in health guidelines, e.g., [11]. The Harvard School of Public Health recently published a new study about the negative impact of CO_2. The study found that high concentrations of CO_2 can reduce decision-making performance [12]. On the other hand, VOCs are emitted as gases from certain solids or liquids that include a variety of chemicals, some of which may have adverse short and long term health effects. Focusing on relative humidity, and referring to [13], it should be remembered that "Relative humidity also affects the rate of gassing of formaldehyde from indoor building materials, the rate of formation of acids and salts from sulphur and nitrogen dioxide, and the rate of formation of ozone". The influence of relative humidity on the abundance of allergens, pathogens, and noxious chemicals suggests that indoor relative humidity levels should be considered as a factor of indoor air quality.

Among studies on monitoring solutions for IAQ, it is worthwhile noting previous theoretical research which was aimed at defining a metadatization approach to develop real-time sensor networks based on diffuse semi-open platforms for to connect and share low-cost Internet of Things (IoT) nodes—e.g., Arduino or RaspberryPi driven ones [14]. This study defines a method to classify and describe in diffused networks the nodes used that allow users to easily include metadata which are able to define the sensor accuracy, errors, monitored variables, etc. Furthermore, a sample application of three nodes in different parts of a city was tested—see also [15]. Nevertheless, commercial applications are now available on the market, such as the Netatmo series of sensors and the related open weathermap service, with which it is possible to access real-time monitored environmental parameters, including outdoor temperatures, relative humidity, atmospheric pressure, and wind direction and velocity [16]. This data source principally shares environmental data. Furthermore, other solutions, including IAQ variables, were developed by the Libelium Company. In particular, the SmartGases Pro system [17] is noteworthy as it involves a plug and sense platform to monitor most air quality parameters including CO, CO_2, O_2, O_3, NO_x, SO_2, NH_x, H_2, ETO, luminosity, relative humidity, atmospheric pressure, and temperature. Each station can act as a node of a network which communicates according to several protocols including Wi-Fi and 4G radio. Other configurations are possible, including smart city ones, e.g., analyzing PM_x levels, noise and other additional variables [18]. Specific applications were also defined in other references, such as [19], in which a modular IoT platform for IAQ is presented in consideration of the mentioned sensor interface board (Gas Pro) while detecting CO_2, CO, SO_2, NO_2, O_3, Cl_2, temperature and relative humidity. These solutions are devoted to air monitoring and are not directly coupled with actuators.

Furthermore, with regard to the connection between IAQ variable level detection and ventilation system activation we should mention the study [20] which refers to a control strategy for demand-controlled ventilation systems (DCV) based on CO_2 detection in multi-zone office spaces. The approach was tested on a very large office building in Hong Kong and works in combination with air handling units (AHU) using an intelligent building management and integration platform. Similarly, in Ref. [21] a study on the automatic activation of an energy recovery ventilation (ERV) system based on a CO_2 detection system was introduced. Generally, ERV are manually controlled, causing potential risks of both insufficient ventilation rates, especially during night time (users' sleeping time), and/or overventilation with consequent heat losses in the cold season. For these reasons, the authors of this study [21] developed a solution to automate the ERV system based on CO_2 levels. An analysis of CO_2 sensor localization in residential apartments was performed, and suggested using the living room to control the entire flat. Another study focuses on the reduction of requested airflow in DCV systems based on CO_2 level detection in indoor spaces to avoid excessive energy consumption when IAQ levels are at the comfort threshold [22]. Furthermore, this research focuses on the usage of simple motor flaps instead of more widely-used variable air volume boxes. Pressure drops are here defined with respect to a developed model and coupled with a CO_2 concentration model, showing a good potentiality in reducing energy consumption, by defining air exchanges in compliance with the model. Another interesting work on a CO_2-based control approach for ventilation systems was proposed in [23]. This reference, unlike other approaches, considers outdoor CO_2 levels as not constant, but fluctuating around daily quasi-periodical variations. In this case, ventilation rates are estimated without directly monitoring CO_2 in the building, showing high-resolution results. Nevertheless, outdoor pollutants may affect the potential of ventilation under certain conditions—see for example [24].

All these studies demonstrate, on the one hand, continuous attention to the development of solutions to control ventilation systems for IAQ purposes, and on the other hand, that the majority of the efforts involve mechanical integration solutions with potentially high installation costs or advanced modelling approaches to avoid sensor usage. Considering residential houses and small offices, only a few commercial solutions adopt IAQ sensing (e.g., CO_2) coupled with ventilation control systems—e.g., detached mechanical ventilation—and include a user interfaces. An example is the mechanical ventilation system Healthbox® produced by RENSON, which connects a multi-zone extractor system with CO_2 zone detection. Inlet air supply is driven naturally in this solution—see [25]. Nevertheless, in the current era of information, with increased access to low-cost sensors the need to develop correlated platforms to integrate sensing with actuating in simple low cost systems is evident, especially if they can be connected to user interfaces to inform and receive feedback.

1.1. Paper's Objectives

This paper focuses on CO_2, Total VOC, and relative humidity, which are some of the main pollutants produced indoors which influence IAQ. Furthermore, the temperature is also detected and used as a risk index to combine IAQ ventilation purposes with ventilative cooling ones, see for example IEA EBC Annex 62 reports [26,27]. In order to act on IAQ levels to maintain acceptable conditions, it is important to have a real-time measurement of the indoor air quality to avoid and prevent related problems in time. The visualization and storage of historical environment data enables understanding and anticipation of current and future situations. Also the use of both environmental and indoor measurements leads to a faster dilution of pollutants thus leading to better comfort levels by knowing air mixing conditions. All these aspects are considered in the proposed system.

The main objectives of this project are as follows:

- to develop an IoT system for to monitor and visualize indoor air quality;
- to adopt a user-friendly approach to present monitored data by using an Indoor Air Quality Index (IAQI) which is able to process different pollutants at the same time and which can be represented on a user-friendly colour scale.
- to carry out a proper ventilation action, by activating a fan-assisted ventilation system, to improve indoor air quality and maintain the desired IAQI level;
- to control fan velocity for IAQI by balancing the external and internal conditions of all pollutants thanks to monitored data;
- to develop devoted user interfaces by using app mobile visualization and data storage for different user profiles (simple users, color scale index; and expert ones, sensor monitored levels);
- to test the IoT system under prototype conditions to reach a Technology Readiness Level (TRL) between 3 to 4.

Additionally, the described approach faces a didactical and methodological issue related to teaching and learning multidisciplinary issues in the building and ICT fields in the context of smart city and smart society problem solving. The Authors come from different backgrounds including telecommunication engineering, data analysis, and environmental and technological building design, under the Interdisciplinary Project activity of the Master Degree in ICT for SmartSociety of the Politecnico di Torino University.

1.2. Paper's Structure

Section 2 of the paper describes the materials and methods adopted to develop the proposed IoT system. In particular, the backend approach is detailed in Section 2.1, the hardware components are introduced in Section 2.2., the mobile app in Section 2.3, and the adopted IAQI index in Section 2.4. Meanwhile, Section 3 is devoted to describing system design and implementation. Section 3.1 focuses on detailing the system architecture, while Sections 3.2 and 3.3 focus respectively on the system's Hardware and Backend definitions. Furthermore, Section 3.4 describes the developed Mobile application and Section 3.5 the developed control strategy. Section 4 focuses on early experimental testing of the system using a prototype based on two-nodes, one outdoor and one indoor and a platform backend which are all connected to the cloud. This section is sub-divided into (i) early validations on single aspects of the platforms (Section 4.1), (ii) prototype box testing considering each pollutant individually under high concentrations, and (iii) continuous prototype box testing producing different pollutants in realistic concentrations. The TRL reached with these tests is between 3 and 4. Finally, Section 5 reports discussions and conclusions, focusing on potential further development steps and details the main outcomes of the paper.

2. Materials and Methods

As shown in Figure 1, this project implements an IoT system to be installed in residential or small office spaces to control IAQ levels based on monitored data. A sample IoT application was also developed to be tested in a small room prototype box. The proposed solution monitors the indoor and outdoor environmental conditions by considering the following parameters: CO_2, VOCs, atmospheric pressure, relative humidity, and temperature. Using these measurements, the IAQI index is computed and the obtained values are used in a ventilation control algorithm to provide the best indoor comfort level. Moreover, a mobile app is developed which provides functionalities of device management, real time measurement observations, and analytics of environmental measurements. The system consists of four main components: Backend, Hardware, Mobile App, and IAQI computation. The following sub-sections are devoted to introducing the methods and materials used for these four components.

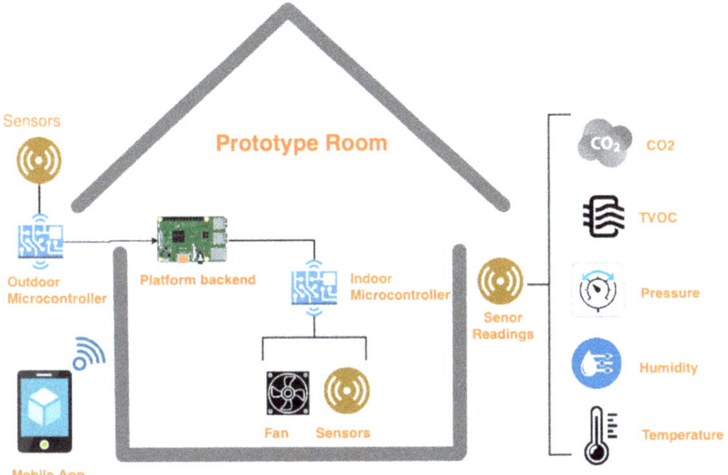

Figure 1. Overview of the developed system, considering the sample test case.

2.1. Backend

The backend should be developed for the distributed internet of things platform in order to handle the sensor data, perform visualization issues, and manage control algorithms. For the purpose of this paper, the considered backend main features are listed here below:

- Device management
- Sensor microcontroller interface
- Sensor data communication
- Control algorithms
- Data storage and aggregation

2.1.1. Design Methodology

The Micro-service architecture approach was selected for the backend design to achieve the objective of a distributed and scalable system. This architecture allows us to structure the proposed application as a collection of loosely coupled services. Modularity makes the application easier to understand, develop, and test. Furthermore, it becomes more resilient to architectural erosion. All the backend features are defined as independent services that can communicate with each other and provide the required functionalities for the other system components.

2.1.2. REST APIs

Using the MicroService design methodology, the system's features are provided as REST Web APIs using HTTP requests. Each Application Programming Interface (API) will provide a feature and can be accessed using HTTP requests to either get, edit or manage operations and data. Python language was chosen to develop the APIs due to the wide availability of required libraries and the simplicity, and scalability of its code. In order to implement the functionalities of the HTTP requests, CherryPy can be used [28]. CherryPy is an object-oriented web application framework which uses the Python programming language. It is designed for the rapid development of web applications by wrapping the HTTP protocol but stays at a low level.

2.1.3. Sensor Data Communication Protocol

Environmental data generated by the sensors are transmitted to calculate the air quality index and to feed control algorithms. Furthermore, the monitored data are also used in the mobile application for real-time data visualization, and they are stored in the database for further aggregations. To achieve these requirements, MQTT communication protocol was used. MQTT is an extremely simple and lightweight publish/subscribe messaging protocol, designed for constrained devices and low-bandwidth, high-latency, or unreliable networks [29]. The design principles of the proposed system are to minimize network bandwidth and device resource requirements whilst also attempting to ensure reliability and some degree of assurance of delivery. These requirements make the protocol MQTT ideal to enable the sensor to publish the data to different subscribing MicroServices in a reliable way.

The implementation of the MQTT protocol requires an MQTT broker, Eclipse Mosquitto Broker [30] was used since it supports MQTT protocol versions 5.0, 3.1.1 and 3.1. It is lightweight and suitable for use on all devices from low power single board computers to full servers. As for the programming of the MQTT clients, Eclipse Paho MQTT library was used to provide all the functions and classes needed to configure a subscriber or publisher application.

2.1.4. Database

The developed IoT platform functionalities require the storage of (i) system configuration; (ii) device information; and (iii) sensor data. MongoDB is a cross-platform document-oriented database program that allows for these functionalities [31]. Classified as a NoSQL database program, MongoDB uses JSON-like documents with schemata. MongoDb supports large volumes of structured, semi-structured, and unstructured data models, allowing us to perform the needed requirements for the proposed IoT application considering data storage functionalities to be used in the micro-services' architecture.

2.2. Hardware

The following hardware components were used for the embedded system and the sensor network.

2.2.1. The Embedded System

RaspberryPi and Arduino. Raspberry Pi—see Figure 2a—is a credit card sized single board computer running Linux Raspbian and is used in the developed IoT system to allocate all microservices and processing power into one stand-alone unit. In particular, the Raspberry Pi 3B+ board was chosen as the master unit to control the operation of the whole system. Moreover, the need for a unit which is compatible with several communication protocols and digital and analog I/O pins has led to the choice of using Arduino Uno R3 shields—see Figure 2b. This specific microcontroller board is compatible with all used sensors and is powerful enough to handle all the acquisition and pre-processing tasks for the slave units, both localized indoor and outdoor.

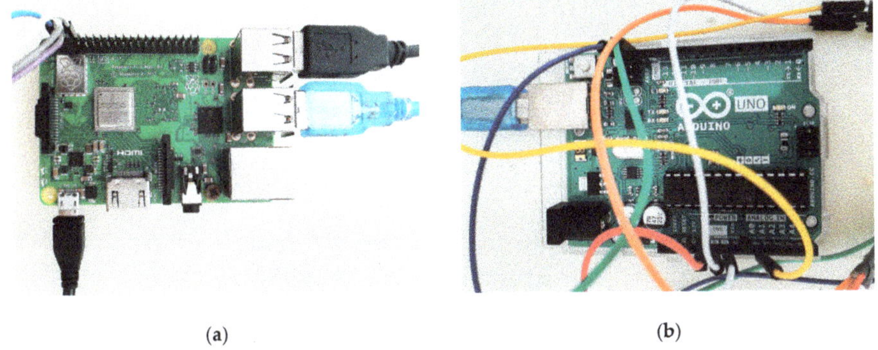

(a)   (b)

Figure 2. (**a**) a view of the master unit used (Raspberry Pi 3B+) (**b**) one the used slave unit board used (Arduino UNO R3).

Ventilation system: Artic P12 PWM—see Figure 3a. The Artic P12 is a PC fan controlled by pulse width modulation (PWM) signal. It has a speed range of 200–1800 RPM and can generate an airflow of 95.65 m^3/h (@1.800 RPM). For the purpose of this paper, this fan was used to simulate the assisted ventilation system implemented in a prototype box.

LCD05—I2C/Serial LCD—see Figure 3b. The LCD05 is a LCD display that adds additional features in comparison with other displays, being fully compatible in both software and pin-out. For the purpose of the developed IoT system, it was used to display useful messages to support the development phases.

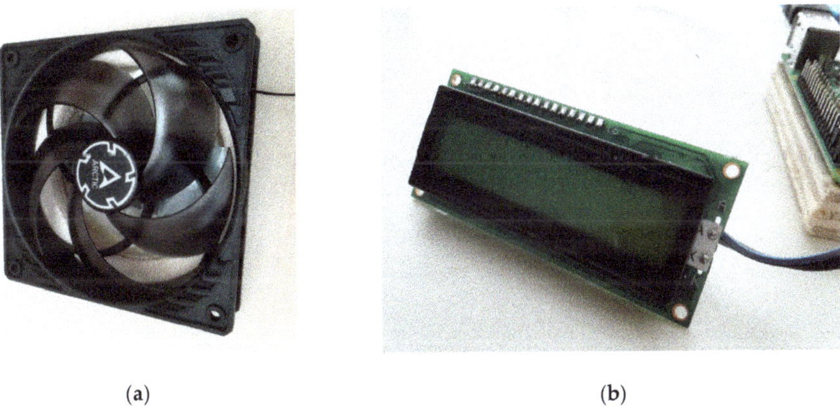

(a)   (b)

Figure 3. (**a**) Ventilation device used to simulate the controlled ventilation system in the prototyping box used(**b**) the LCD display screen used for development phases.

2.2.2. Sensors

BME680. The DFRobot BME680 Environmental Sensor (DFRobot, Shanghai, China)—see Figure 4a—is a lowpower sensor based on the BOSCH BME680 (Bosch Sensortec, Reutlinger, Germany). It is a 4-in-1 environmental sensor, which integrates a total VOC (Volatile Organic Compounds) sensor, a temperature sensor, a relative humidity sensor, and a barometric pressure sensor. This specific hardware component is characterized by a large availability of libraries which facilitate the acquisition of data and the development of the specific code. Pressure absolute accuracy is ±0.6 hPa, relative humidity one ±3%, temperature ±0.5 °C (at 25 °C), ±1 °C (full range), VOC 2–5% depending on specific VOCs [32].

(a)   (b)

Figure 4. (**a**) the BME680 Temperature, Relative Humidity, Barometric pressure, and TVOC (Total Volatile Organic Compounds) gas sensor used; (**b**) the used Gravity Analog CO_2 gas sensor used.

The CO_2 Gas Sensors used are illustrated in Figure 4b. Two of them are used to detect the CO_2 levels of the indoor air and of the environmental air. This specific device is based on the MG-811 gas sensor (DFRobot, Shanghai, China), which shows high sensitivity to CO_2 levels and less sensitivity to alcohol and CO presence. It is characterized by low relative humidity and temperature dependencies. The industrial quality of this component makes it sufficiently robust and stable despite its low-cost. Moreover, it is characterized by a measurement range of 350–10000 ppm, while its accuracy is within the range of 1% of the measurement scale. The proposed system is, however can also be connected to different sensors, such as the analogy infrared CO_2 Gravity sensor for Arduino, which is more accurate (50 ppm + 3% reading).

2.3. Mobile App

An Android application was designed to allow users to monitor and control the system. Two main user profiles were adopted: on the one hand, users with limited technical knowledge of IAQ indexes and parameters; on the other hand, operational users with specific technical backgrounds. The former access to a colored graphical interface which classifies the IAQI value for each parameter on a scale based on four classes—see the following Section 2.4. For the latter the app reports the monitored values for each variable by considering temperature [°C], relative humidity [%], atmospheric pressure [hPa], CO_2 levels [ppm], and TVOC (Total Volatile Organic Compounds) concentration [ppm]. Both user profiles may have access to indoor and outdoor data, even if for the non-technical profile indoor conditions are of most interest.

The main features of the app are listed below.

- Device management: it allows us to create a list of the installed devices for each specific application (e.g., a specific building, a specific zone) and to set a desired ventilation time to report indoor conditions in threshold value for each device;
- Real time data visualization: data from all sensors of a connected device are visualized on a single screen, thus permitting the users to always have an overview of the quality of the air, be they expert or non-expert users
- Statistical data: daily or weekly data graphs for each sensor are shown for both user profiles.

Figure 5 shows the developed mobile app navigation flow between main app components.

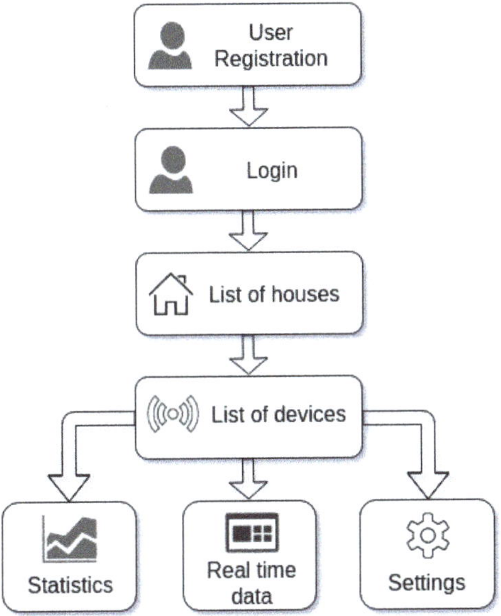

Figure 5. Mobile app navigation map.

2.4. Indoor Air Quality Index Computation

Considering the variables monitored in the proposed system, each has its own specific unit of measurement. Hence, relative humidity is measured in percentages [%], temperature is measured on the Celsius scale [°C], the concentration of CO_2 and of TVOC are converted into parts per million [ppm]. If these values can be easily read by the expert user profile, the non-expert user may have difficulty in understanding these results. For instance, if the concentration of CO_2 is 700 ppm, it's hard to understand if the air quality is good or not when users are not experts on specific standards. For this reason, it is necessary to make the measurements more readable and understandable for people who do not have specific background knowledge. To face this challenge, the developed system labels the concentration of CO_2 and of the other considered variables by using intuitive categories (good, bad, very bad, etc.) in line with the Air Quality Index (AQI) approach. This index was developed by the United States Environmental Protection Agency (US EPA) in 2006 focusing on outdoor conditions [33]. Nevertheless, an indoor version of this index was recently presented [34] and is here adopted to compute the IAQ level. The AQI approach converts measurements of different pollutants in consideration of their specific unit of measurement into the AQI code—see for example [33,35]. In particular, for the purpose of this study, the break point table for indoor environment IAQI classification reported in [34] was adopted—see Table 1. IAQI is, in fact, a numerical classification scale ranging from 100 (perfect condition) to 0 (worst condition) and also a color code, which divides the numerical classification into several specific ranges (good, moderate, unhealthy, hazardous). AQI index describes not only the pollution level but also the potential risk level for people's health, such as was illustrated in [33]. The adopted threshold values were defined in accordance with reference standards [34]. For example, for carbon dioxide, the same ranges are proposed by [36], while 1000 ppm is assumed as a limit by [37], which also introduces a lower class for optimal (below 800 ppm), by [38], and is used as reference by [21]. ASHRAE standard 62-1989 also consider 1000 ppm as target concentration level even if it does not consider this threshold as a standard requirement—see also [39]. Furthermore, 1500 ppm is a breakpoint for schools in the UK [40]. Nevertheless, potential variations can be considered for local standard adaptations or to adopt specific thresholds. For example, the temperature thresholds can follow an adaptive comfort approach when the system works in non-conditioned buildings—see for example [41]. In addition, the IAQ standard EN 13,779 classifies the CO_2-related IAQ level into four classes while considering a carbon dioxide concentration which is above that of the outdoor air (i) less than 400 ppm, (ii) in the range 400–600 ppm, (iii) between 600–1000 ppm, and (iv) above 1000. The same standard suggests typical CO_2 outdoor air ranging from 350 ppm in rural areas to 450 ppm in city centres. AS is illustrated in Table 1, the IAQI values and status are divided into four categories: Good (100–76), Moderate (75–51), Unhealthy (50–26) and Hazardous (25–0). Depending on different IAQI values for each pollutant, proper control strategies will be adopted to improve IAQ.

Table 1. Classification scale adopted for relative humidity, temperature, CO_2, and TVOC detected levels. The considered thresholds are based on [34], but are in line with reference standards. Potential variations due to local standards may be included by varying specific variable ranges acting on the user interface setting function.

CO_2 [ppm]	VOC [ppm]	Temperature [°C]	Relative Humidity [%]	IAQI Index	IAQI Status
340–600	0.000–0.087	20.0–26.0	40.0–70.0	100–76	Good
601–1000	0.088–0.261	26.1–29.0	70.1–80.0	75–51	Moderate
1001–1500	0.262–0.430	29.1–39.1	80.1–90.0	51–26	Unhealthy
1500–5000	0.431–3000	39.1–45.0	90.1–100.0	25–0	Hazardous

Formulas to Support System Control and Analysis

The proposed IoT system uses simple well-known expressions to support the decision process of the algorithm. In order to define the proper action to control the ventilation system, two main parameters are considered: the time span chosen to reach a sufficient IAQ level (ventilation system operation time) and the airflow rate. Users will directly set the optimal time threshold to reach a good IAQ level by activating the ventilative system and thus releasing pollutants from the inner environment. For this reason, only the airflow rate has to be set automatically by the platform. Firstly, the ventilation rate Q [m^3/s] needs to be calculated. Referring to [21–23,42], the following expressions—see Equations (1) and (2)—were adopted to calculate the required Q to obtain, within the set time, sufficient pollutant dilution.

$$(Q \cdot t) \cdot \text{var}_{out} + V \cdot \text{var}_{in} = (V + Q \cdot t) \cdot \text{var}_{final} \tag{1}$$

where t is the ventilation time [s], V is the zone volume (e.g., the prototype box) [m^3], Q is the ventilation rate [m^3/s], and *var* is the considered variable (temperature [°C] or relative humidity [%]). Subscripts $_{in}$ refers to current indoor environmental conditions, and $_{out}$ refers to current environmental (outdoor) conditions, while $_{final}$ refers to the final required condition.

$$C_{final} = C_{out} + (C_{in} - C_{out}) \cdot e^{-\left(\frac{Q \cdot t}{V}\right)} + \frac{G}{Q} \cdot \left(1 - e^{-\left(\frac{Q \cdot t}{V}\right)}\right) \tag{2}$$

where C is the concentration value of the considered pollutant (TVOC or CO_2) expressed in [ppm], G is the indoor generation rate of pollutant [m^3/h], Q the ventilation rate in [m^3/h], and t the ventilation time in [h]. Furthermore, V is the zone volume [m^3], while subscripts are the same as in Equation (1). Nevertheless, the pollutant emission rate is not commonly known, especially if human occupancy is not fixed or monitored. For this reason, for the sample prototype box, the generation rate of CO_2 is set to zero, and the generating rate G for VOC is negligible, fixed emission sources not being present. The adopted testing approach focuses on dilution of high level of quickly produced pollutants to demonstrate the effect of the system under stressful conditions rather than under the continuous production of a small amount of them, such as under fixed human occupancy profiles where CO_2 production can be predicted. In fact, when pollutants are placed inside the prototype box, the generation rate will vary all the time, and it is hard to calculate the correct value, unlike under real conditions. In this case, the generation rate is not considered for further analysis. On the other hand, calculations are based on conditions where indoor and outdoor air mix to reach the desired indoor IAQ in the given time.

For temperature, CO_2 and TVOC, the required ventilation rate Q can be calculated directly by applying Equations (1) and (2). Nevertheless, for relative humidity (*RH*), the measurements should be converted before performing calculations because *RH* is a function of temperature, atmospheric pressure, and moisture content (X) [kg/kg]. For this reason, indoor and outdoor measured values cannot be directly compared to define the airflow needed to reach the desired threshold, considering that indoor temperature and/or moisture content could be different from those outdoors. Relative humidity is defined as the ratio of the water vapour pressure (P_w) [Pa] to the saturation water vapor pressure (P_{ws}) [Pa] at a given temperature:

$$RH = \left(\frac{P_w}{P_{ws}}\right) \cdot 100 \tag{3}$$

Psychrometric expressions are used to define saturation water vapour, water vapour, and moisture content. Cetiat tables of the humid air or other comparable approaches may be used—see for example [43].

$$P_{ws} = EXP(((A \cdot \vartheta)/(B + \vartheta)) + C) \tag{4}$$

where ϑ is the monitored air temperature [°C], while A, B, and C are constants. In the $-0\,°C\ldots +40\,°C$ domain A, B and C were assumed to be respectively 17.438, 239.78, and 6.4147, in line with Cetiat suggestions [43]. The moisture content may be calculated according to the well-known expression (5):

$$X = 0.662 \cdot \frac{RH \cdot P_{ws}}{P_{air} - RH \cdot P_{ws}} \tag{5}$$

The reverse formula (6) allows us to estimate the *RH* by knowing the air moisture content and the air temperature. This expression can be used to roughly estimate the *RH* that the outlet air would reach if sensibility treated, such as under the effect of high efficiency heat recovery systems, like the ones installed in passivehauses.

$$RH = \frac{X \cdot P_{air}}{0.622 \cdot P_{ws} + X \cdot P_{ws}} \tag{6}$$

Then the required ventilation rate for relative humidity is calculated through Equation (1) by using indoor measurement and the converted outdoor measurement. According to these formulas, the required ventilation rate for each pollutant can be calculated within the expected ventilation time.

According to the required ventilation rate Q [m³/s], it is possible to define the required air velocity, q_{fan}, to be performed by the fan [m/s]. This can be calculated by applying the following expression (7):

$$q_{fan} = Q/A_{fan} \tag{7}$$

where A_{fan} is the net opening area of the fan [m²].

To control the fan speed, different PWM signals should be fed to the device (fan). Pulse width modulation (PWM) is a method which reduces the average power delivered by an electrical signal. The delivered average power is controlled by changing the duration of the pulse, which leads to different duty cycles. The higher the duty cycle, the larger stronger the average power [44]. In order to feed a proper PWM signal to the fan, the relationship between the set PWM and the produced air velocity is very important. An anemometer was employed to correlate the produced air velocity of the fan with different PWM signals in the prototype box. Figure 6 shows the values obtained, while the conversion between the fan air velocity and PWM signal is achieved based on the plotted graph.

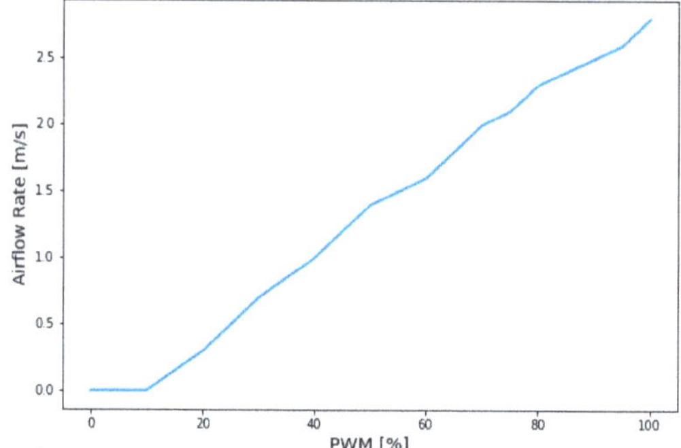

Figure 6. The relation between pulse width modulation (PWM) and air velocity measured in the prototype box.

3. Design and Implementation

3.1. System Architecture

The developed IoT system consists of four main levels as is shown in Figure 7. The User Interface level is represented by the mobile application that manages all devices and configurations, and furthermore displays sensor data. The Backend level provides functionalities of data storage, APIs for platform management, analytics, and real time sensor measurement storage. Meanwhile, the gateway level is responsible for controlling and accessing the hardware level and connecting it with the backend. On this level, control strategies are implemented to process the environmental data and apply algorithms to calculate the IAQI index and control fan operation.

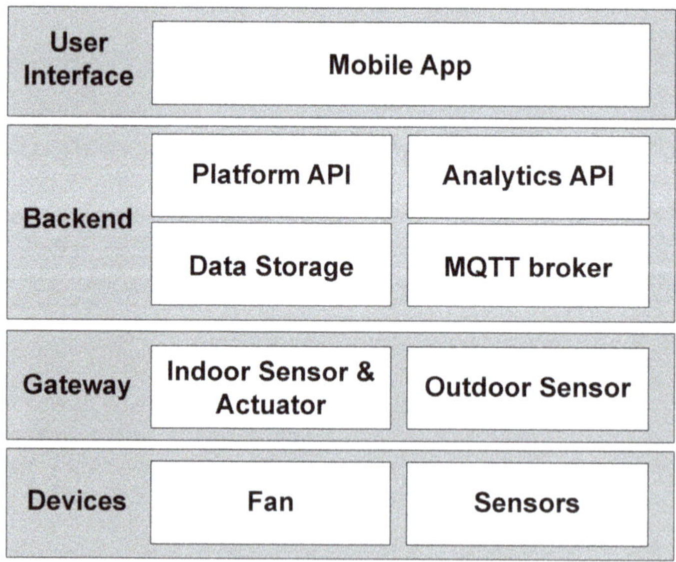

Figure 7. The four defined Backend levels.

3.2. Hardware

3.2.1. Wired and Wireless Connections

For the main unit (Raspberry Pi 3B+), the most used functionalities were the GPIO pins, WiFi, and USB ports. The GPIO pins supporting I2C protocol are used to connect the board with a LCD screen, the screen is employed to show actual indoor conditions (Figure 8c). As regards the software, different Python scripts are run on the board to manage, monitor, control, process, and store data. Moreover, Arduinos are used to collect and pre-process the raw data from sensors while the processed data are further transmitted to the master unit (the Raspberry Pi) through serial communications. Outdoor unit connections are shown in Figure 8a. Through a pin socket, the fan is supplied up to 12 V DC by an external source and is controlled by one of the PWM pins of the indoor Arduino. It reacts to the master's order by adjusting the speed signal. The deployment of sensors, fan, and Arduino board for an indoor unit is shown below in Figure 8b.

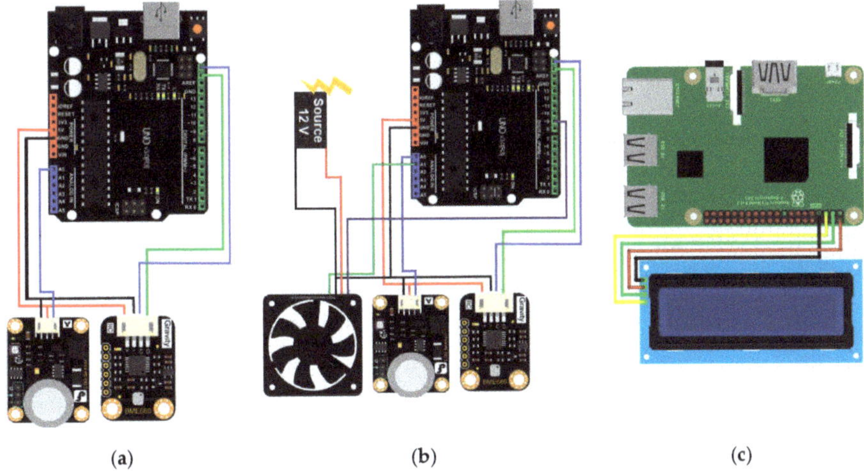

(a)   (b)   (c)

Figure 8. (**a**) Outdoor unit hardware connections; (**b**) indoor unit hardware connections; (**c**) raspberry Pi connection (master unit).

The WiFi functionality of Raspberry Pi is used to create a hotspot to connect microservices with an external device which runs the mobile application. The application is used to show the state of all the environmental conditions (indoor and outdoor) and to set up the user parameters for the operation of the system.

3.2.2. Working Flowcharts of the Indoor and the Outdoor Units

Figure 9 shows the working flowchart of indoor and outdoor units for raw data acquisition, data pre-processing, and serial communication. Indoor and outdoor units will pre-process sensor readings and transmit these data to the master unit. In the case of the indoor unit, the master unit (Raspberry Pi) sends back the control signal (PWM) in order to achieve the target established by the control strategy (implemented on the master unit) to maintain the indoor air quality at a good level.

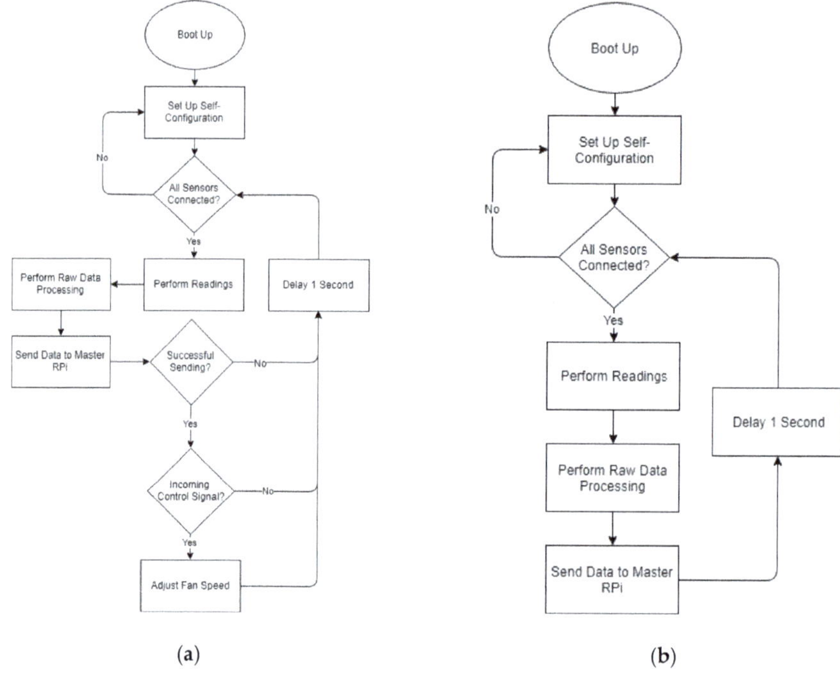

Figure 9. Embedded software flowcharts (**a**) Indoor unit processing flowchart; (**b**) outdoor unit processing flowchart.

3.3. Backend

The backend follows the Microservice architecture—see the backend overview in Figure 10. Therefore, an initial service "Resources/Services catalog" was implemented in order to manage events and data flow between microservices. Initially, each microservice should contact the Resources/Services catalog to get the list of available microservices and find the way to access these resources. The communication between microservices happens through REST and MQTT. As for the databases, the communication is managed through the Mongodb communication tunnel.

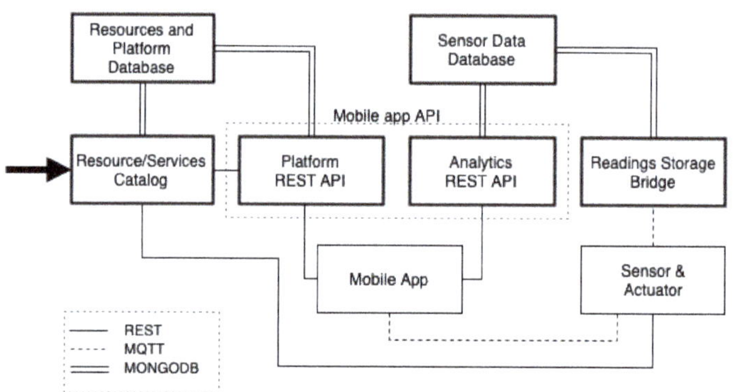

Figure 10. Backend overview.

3.3.1. Database

The developed system consists of two independent databases based on Mongodb. Each database can be accessed independently through the Mongodb python connector from another microservice. The connection is guaranteed by using TCP on the port 27,017 and by specifying the installation IP address of the database. The two databases are described here below.

Resources and Platform Database: this database is used to store information about platform services and resources such as mobile application user management, devices, configuration profiles, and available services. The database is schematised in Figure 11. This figure shows the relations and the structure of each collection process. The *house* collection stores the information about the house/room in which the sensors are installed. The *services'* collection stores information about the available system microservices and the related addresses for contact. *Profile* collection contains predefined room profile(s) of thresholds that provide the best comfort levels. *Sensor settings'* collection stores the device configuration(s) and threshold(s) for a selected sensor. *Device* collection stores the information about a device how to access its resources. Finally, the *sensor_house* and *user_house* collections are responsible for defining the parent data collection to which the entity belongs.

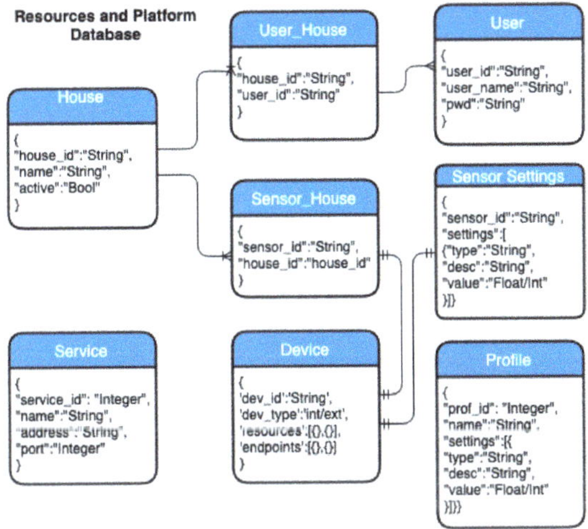

Figure 11. Resources and Platform database schematic view.

Sensor Data database: this database is used to store all sensor measurements inside the platform. It consists of only one collection called "data". In this collection, the sensor measurements are stored in a consistent format as shown in Figure 12. The "bn", "bt", "n", "u", "v" expressions define the sensor id, time of measurement, type, unit and value. This format makes the platform more amenable to the introduction of new sensor types without impacting on the database structure.

```
{
    "_id" : ObjectId("5c8b7b8eae5098231a798fa5"),
    "bn" : "SS1",
    "bt" : ISODate("2019-03-15T10:16:46.000Z"),
    "n" : "iaq",
    "u" : "%",
    "v" : 0.202906692557987
}
```

Figure 12. Measurements data structure.

3.3.2. Microservices

Microservices are python scripts which run independently and they can be accessed using REST calls provided by cherrypy. In particular, the following list of microservices was implemented.

- Resource and service catalog: the resource and service catalog microservice is the entry point to the system. It mainly provides the functionalities of device management as well as the list of services.
- Readings storage bridge: the main functionality of this microservice is to preprocess the incoming sensor measurement messages and store them in the *sensor_readings* database. It subscribes MQTT topic "Readings/#", loads the incoming messages, pre-processes data into the proper format, and stores them in the database.
- Analytics API: this microservice provides sensor measurements from the database to the mobile app. The measurements are filtered by sensor type and aggregated by week or day. Instead of sending all the sensor measurements, the microservice allows us to access average aggregations to return average data value per hour or per minute of the day.
- Platform API: this microservice provides the main functionalities of the mobile app for user, house, and device management.
- Sensor and actuator: on the one hand, the sensor device is running a microservice that reads the measurement from the sensor and publishes it through MQTT to be stored in the database. On the other hand, this microservice requests the configurations and settings from the resource/service catalog.

3.4. Mobile Application

Android Studio IDE is used to develop the mobile application using Java programming language. The application communicates with the backend through a REST API and receives data from the devices through MQTT protocol as it is connected to Raspberry Pi hotspot.

To implement these communications, two libraries were used: *Android Asynchronous HTTP Client* for REST requests and *Paho Java Client* for MQTT communication. Furthermore, *MP Android Chart* library was used for the visualization of the graphs. Furthermore, the list of all mobile application functionalities is given here below.

- Service Catalog setting: during the first access to the app, users must insert the IP address of service catalog, so that the IP address of the platform can be retrieved and the app can connect to this platform. It is possible to access this setting at any time both from the list of houses and from the device list.
- Registration: Users can register themselves by defining username and password.
- Login: users may log into their system. If correct credentials are inserted, the list of profile-connected houses is retrieved.
- House management: the list of houses is the home page of the app. Users can register a new house or access an existing house (based on house id) in their lists.
- Device management: both internal and external devices are listed on the same screen. There are specific buttons to add a new device, to show statistics, and to change settings. During the registration of a device, a room profile (e.g., Kitchen, Bathroom, etc...) based on those available must be chosen. By clicking on an element in the list, the real time data page is visualized.
- Real-Time data visualization: Users can always monitor the quality of the air for both indoor and outdoor sensors. All the readings of a device are shown on a single screen, and different colors give an indication of the pollution levels in the air. This part of the app is based on two user typologies, as mentioned above in the paper. While non-expert users will access a coloured visualization screen which indicates current IAQI status, an expert user interface will show monitored data in consideration of the specific data in the devoted unit of measurement (e.g., ppm for CO_2 and

TVOC; hPa for pressure; °C for temperature, % for *RH*). To simplify these, two user profiles are shown one over the other in the current version of the mobile application—see Figure 13a.

- Statistics visualization: statistics for daily or weekly measurements can also be visualized. A spinner menu allows users to choose the specific measurement to be analyzed. See for example Figure 13b.
- Logout: from the main page or from the device page, users can logout from the system and get back to the login page.

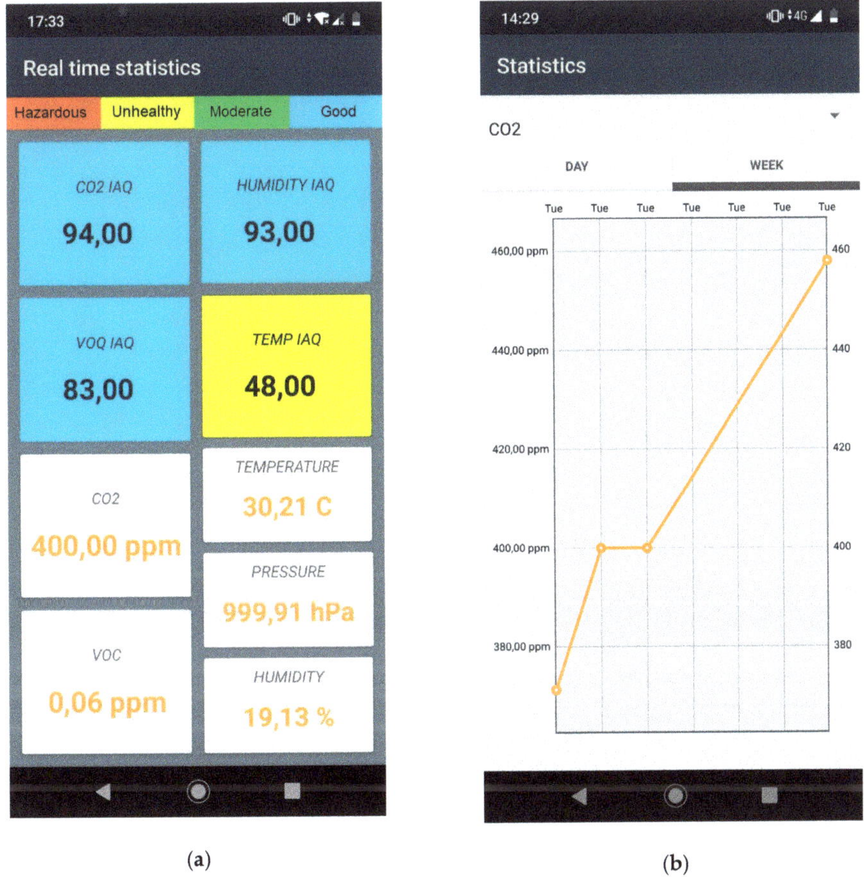

(a)   (b)

Figure 13. Mobile application screenshots: (**a**) real-time visualization, at the top the not-expert user view, at the bottom the expert user view; (**b**) Statistics visualization.

3.5. Control Strategy

Depending on different pollution levels, indoor air quality (IAQ) can be improved by activating the mechanical ventilation system to let external air enter the living space. Ventilation time is quite an important parameter to be taken into account for ventilation activity. In fact, it not only affects the efficiency, but also the comfort level of people. If the required ventilation time is very short, the ventilation system could produce strong airflow velocity, which may not be comfortable for people. On the contrary, if the ventilation time is too long, the efficiency of the ventilation system may not be enough to guarantee sufficient pollutant dilution and to avoid stagnation and high pollution exposure. In the proposed system, the expected ventilation time can be set by users through the mobile application, while the fan speed is regulated dynamically with respect to different monitored pollution levels and the set ventilation time.

Figure 14 illustrates the ventilation decision process which compares the IAQI indexes of indoor and the outdoor air. If ventilation activity is computed to be necessary (e.g., the prototype box shows a high level of one/more pollutant/s), the calculation process will be activated in order to find the proper ventilation rate in order to set the fan system and dilute pollutants. According to expressions listed in Section 2.4., the system will compute the required ventilation rate for each pollutant. When the system gets different ventilation rates for different pollutants, it always selects the maximum ventilation rate from among these values to prevent insufficient ventilation. According to the selected ventilation rate Q [m^3/s], the system calculates the airflow rate [m/s], and then the desired PWM value for fan activation is derived from the corresponding airflow rate by using room characterization—see Figure 6 for the considered prototype box. Finally, the PWM signal is transmitted to indoor units through serial communication to control the action of the fan.

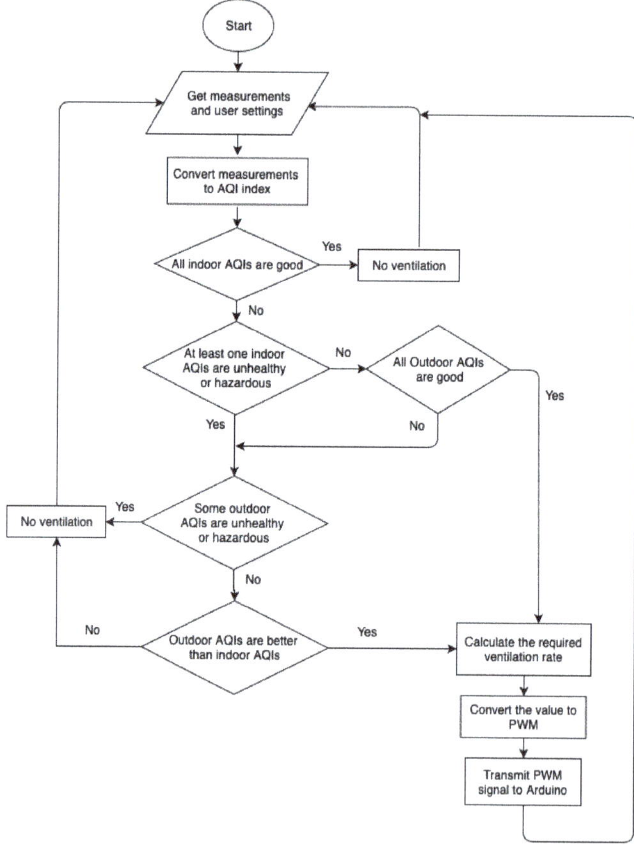

Figure 14. Flowchart of the analysis algorithm.

4. Experimental Functioning of the System (TRL 3–4)

This section focuses on proving the feasibility of the developed IoT systems, with special regard to (i) testing single component' functionalities, (ii) system early-validation in a prototype box which tests the system under high-pollutant concentration condition, and (iii) Prototype box test to check the system's ability to constantly deal with a series of uncomfortable IAQI conditions in consideration of stressful situations that may arise in buildings, with special regard to residential and small office spaces with no mechanical ventilation systems. All of these tests are conceived to reach a TRL (Technology readiness levels) of 3 to 4, representing the first important step in order to plan for further development of the system.

4.1. Single Components' Testing

To ensure the proper functioning of each component of the IoT system, several test cases were performed individually for each part. Each test focuses, in fact, on a specific step. The performed test cases are shown in Table 2. After these specific tests, the components were integrated together to check the behavior of each part when included in the whole system.

Table 2. Test cases for single components' testing.

No.	Test Case	Expected Results	Pass/Fail
1	Hotspot and bootup scripts	Hotspot and system scripts start automatically	Pass
2	Automatic restarting in case of hardware failure	The hardware scripts should return if hardware is reconnected	Pass
3	Storage of sensor data in the database	The sensor readings should appear on the database	Pass
4	Real time sensor reading on the mobile app	The sensor readings should be received through MQTT on the mobile app and be displayed accordingly	Pass
5	CO_2 sensor reading	Test on different environmental conditions	Pass
6	VOC sensor reading	Test on different environmental conditions	Pass
7	Humidity sensor reading	Test on different environmental conditions	Pass
8	Temperature sensor reading	Test on different environmental conditions	Pass
9	Fan actuation	Fan should turn on with the correct speed from the actuation function (tested for all pollutants independently)	Pass
10	External unhealthy conditions stop fan activation	The fan should turn off and prevent pollutants immission	Pass

Firstly, sensors were connected to the Arduinos, and then the sensibility and the performance of each Arduino slave unit was tested with respect to different pollutants. In parallel, the script running on the master unit was probed. The master unit was connected to a laptop to check and correct the issues when they appeared. Furthermore, a wireless connection was created between the mobile application and the master unit to check both outputs of the stand-alone script, and the performance of the application once it got the data.

Following these implementations and tests, all the parts were connected together to perform the final test before deployment. For example, during this phase, one unit was found to be sending incorrect data to the master unit when it was plugged in, due possibly to a production problem in the analog input connectors. The problem was solved by substituting the board.

4.2. Sistem Tests Using a Prototype Box under High-Pollutant Concentrationsusing

After solving these issues, the whole system was deployed in the prototype box to test its efficiency in pollutant dilution when different stress conditions were simulated (e.g., release of TVOC, increase in *RH*%). The prototype box used for these tests was built in plywood (1.5 cm) and insulated with polystyrene panes (3 cm) which were positioned externally all along the box surface—see Figure 15. The ventilation system is made up of the extractor fan and an inlet mouth with the same net opening area positioned at the bottom of the façade opposite. The net box dimensions are $57 \times 60 \times 120$ cm. Finally, a polycarbonate permanently sealed window for visual inspection is installed (29×39 cm).

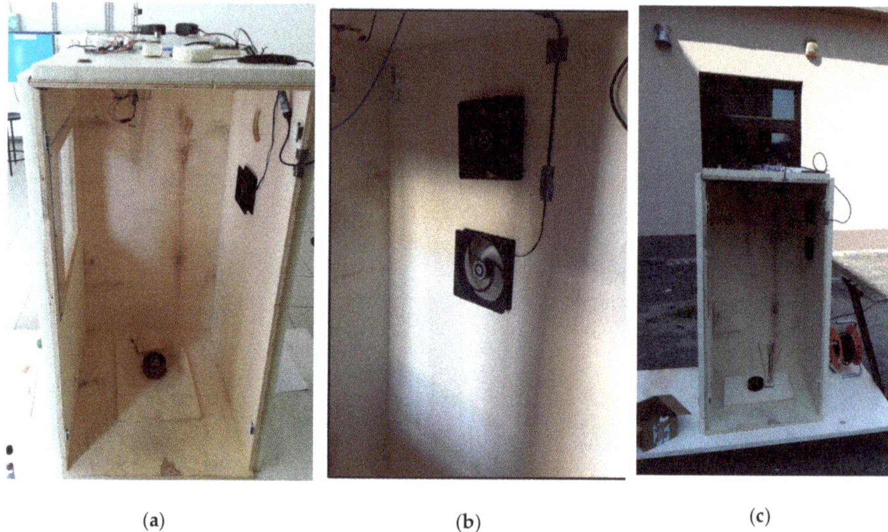

<table>
<tr><td>(a)</td><td>(b)</td><td>(c)</td></tr>
</table>

Figure 15. (**a**) The prototype box (note that to take this picture the inspection-side was removed); (**b**) box interior view (detail); (**c**) the prototype box in the testing location with inspection-side removed.

In Figure 16a the system was tested for relative humidity variations. For this test the prototype box was moved to an external courtyard in a central urban area (North of Italy) position—see Figure 15c. Outdoor conditions were monitored to be 406 ppm of CO_2, 0.03 ppm of VOC, 27.2% of *RH*, and 26.4 °C. The indoor *RH* level was stressed by boiling a kettle. When the value of relative humidity exceeds the threshold of "good level" for a given amount of time (activation risk time threshold e.g., 30 s), the assisted ventilation system starts working to get back to the "good" level according to the related IAQI index. The test demonstrates that the threshold was reached again within about 150 s, which was 30 s quicker than the expected time.

Then the system was tested under uncomfortable VOC concentration conditions. Very high VOC concentrations were produced by burning a large amount of incense in the prototype box. AS illustrated in Figure 16b, once the ventilation system was activated, the ventilation time to reach the "moderate" IAQI level of VOC was approximately 125 s. Because the fan used was not powerful enough to provide the required level of airflow rate to reach the "good" level under uncomfortable VOC concentration conditions, the system could not bring the level of VOC to a good level within the considered activation time (180 s). Nevertheless, as is illustrated in the Figure, the 'moderate' level is reached in the desired time, thus guaranteeing a healthy level of VOC gases.

Both tests showed the correct functioning of the system under typical conditions of high concentration of VOC gases and high relative humidity. The CO_2 test was not performed in this phase since CO_2 and VOC are treated in a similar manner by the IoT system.

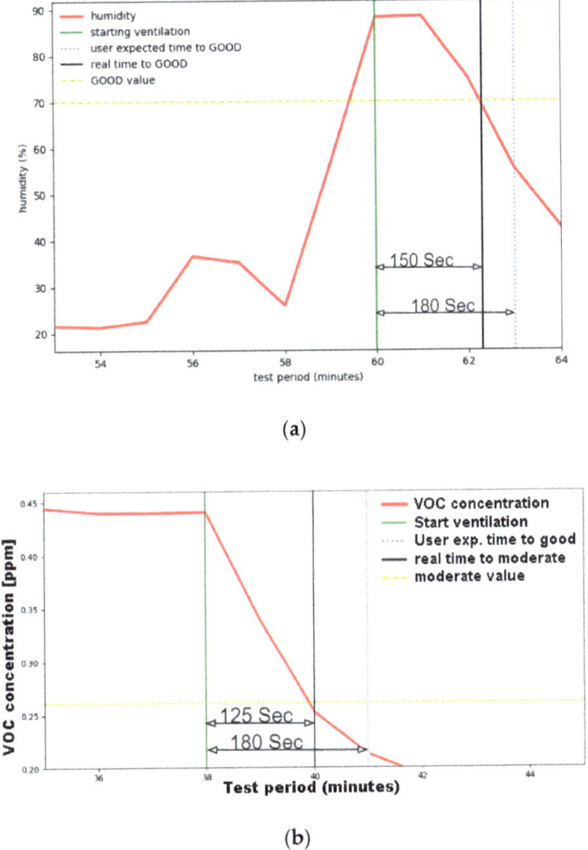

(a)

(b)

Figure 16. Pollution tests: (**a**) relative humidity evolution over time; (**b**) Volatile Organic Compounds (VOC) gases concentration over time.

4.3. Prototype Box Continuous Test (Realistic Conditions)

Once the system was completely checked, realistic test cases were performed. During this demonstration phase, different domestic potential conditions were reproduced in the prototype box. To demonstrate the impact of the IoT system, its behavior was shown over a test time of 1 h and during this period several pollutants were placed inside the box to simulate real-life situations, such as a CO_2 concentration of about 1000 ppm (a value often found in an occupied room after a period of system deactivation—e.g., a double room with two adults in less than 1 h—adult CO_2 production rate of 0.0052 L/s, ASHRAE Standard 62.1 and ASTM D6245 [45]), or a *RH*% higher than 85% coupled with temperature increase—as may arise in a bathroom after a hot shower. In particular, three pollutants were released in sequence: CO_2, VOC, and *RH*. VOC and *RH* are produced in the same way as in the tests for high concentration dilution (incense and kettle), while CO_2 was generated by burning acetone inside the box. During these tests lower levels of pollution were produced in comparison to high-concentration tests.

Figure 17 shows the results of the monitored indoor conditions expressed in terms of IAQI levels during the entire testing period. Each line indicates the IAQI of a specific pollutant, and each color band represents a different IAQI level based on the classification reported in Table 1. During this test, external measured conditions were almost stable with a CO_2 concentration of 400 ppm, a VOC concentration of 0.02 ppm, pressure of 999.73 hPa, a *RH%* of 24.8%, and a temperature of 31.3 °C. With the exception of temperature, all other outdoor variables reached a good IAQI level.

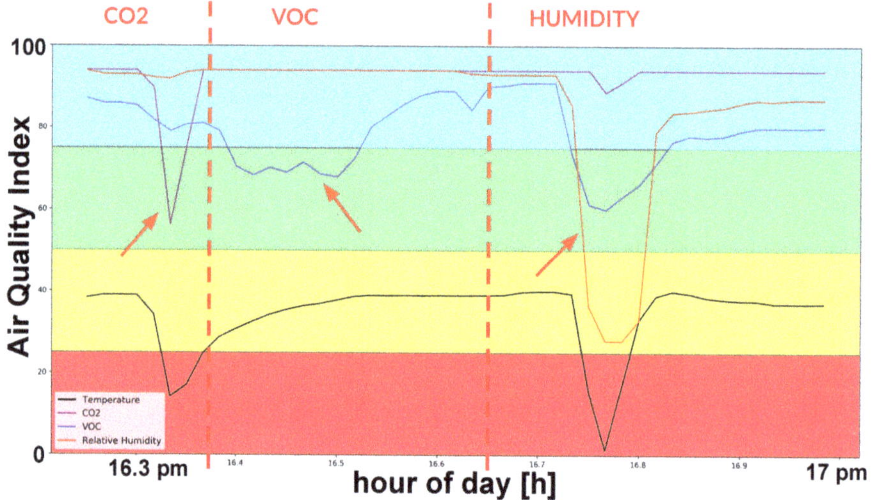

Figure 17. System evolution under different pollutants.

Firstly, at approximately 4:10 PM (16.15 h), acetone was burned to simulate a potential CO_2 concentration of about 1000 ppm. The observed result (Purple Line—IAQI CO_2) was an increase in the concentration of CO_2 which led to lower IAQI. Once the ventilation system was automatically turned on, the concentration of CO_2 decreased drastically. The prototype box was prevented from reaching unhealthy levels of this specific pollutant. The temperature curve (Turquoise Line) showed the impact of the system though this test does not include IAQI temperature control, but rather focuses on on IAQ variables for fan activation.

Secondly, VOCs were produced inside the prototype box (around at 4:25 pm, 16.4 h) to simulate the situation that can be experienced when using some personal hygiene products or simply burning incense. The VOC curve (Blue Line—IAQI VOC) showed that this pollutant concentration did not reach an unhealthy level. However, once the fan was turned on, the VOC level improved quickly from moderate to good. The performance of the system was then satisfactory

Finally, the system was tested by placing a humidifier to simulate the conditions of a bathroom during and after a shower with hot water (see approximately at 4:40 pm, 16.7 h). As is to be expected the IAQI of relative humidity decreases when humidity is produced (Orange Line—IAQI Humidity) thus activating the fan which is controlled by the system to bring this index back again to a good IAQI Level. However, the IAQI of VOC is also reduced. In this specific case, this the increase in VOC is thought to be related to the previous VOC emission the wood in the internal walls of the prototype box may have absorbed a portion of the contaminants previously presented in the air, and, with the increase in water vapor, these are probably released. Nevertheless, in spite of being subject to all different pollutants independently, the system acted in the expected way. The levels of all pollutants were controlled through effective ventilation to provide a healthy indoor environment by correctly setting the fan operation.

Figure 18 shows some pictures taken during this experiment and illustrates both the monitored values on the app and the release of pollution inside the box.

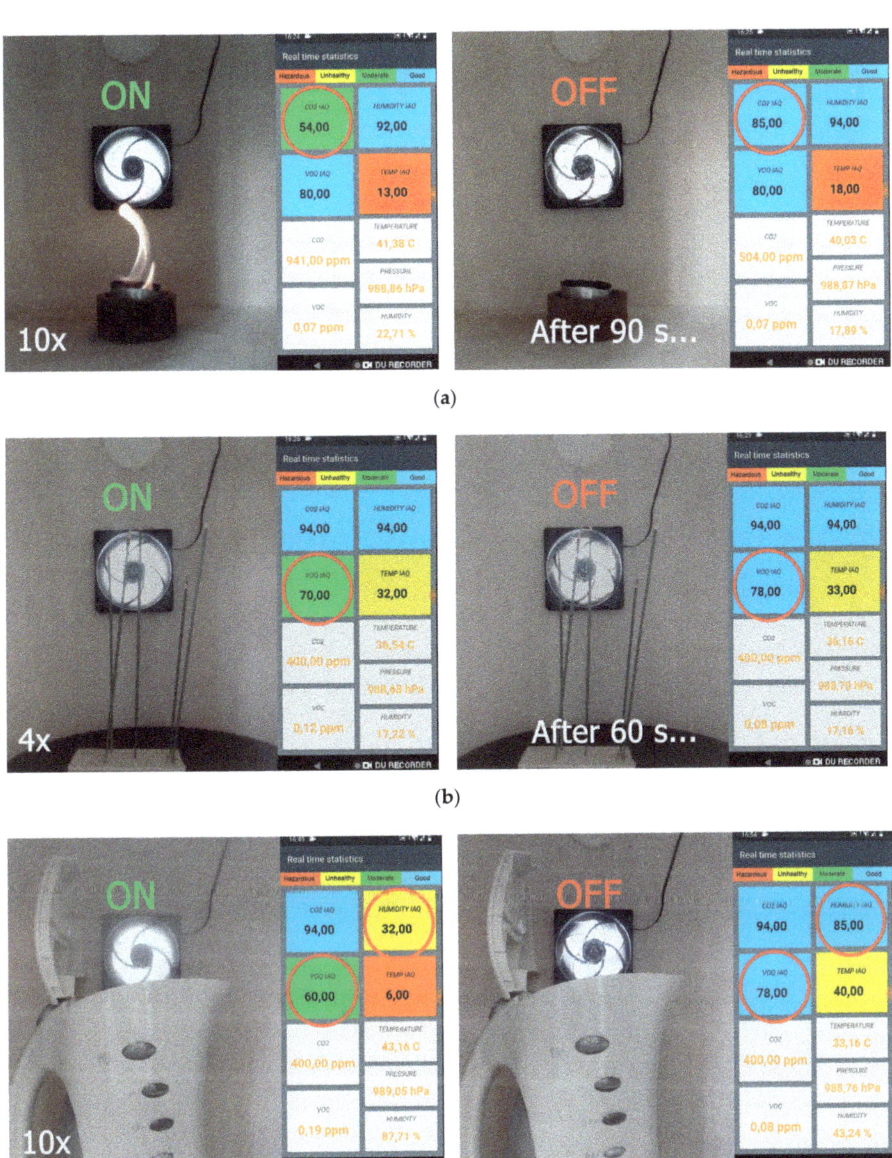

Figure 18. Test conditions and related monitored level of Indoor Air Quality Index (IAQI) on the developed app at the beginning of fan activation and immediately after it being turned off considering (**a**) CO_2; (**b**) VOC; and (**c**) *RH*. Picture are taken though the inspection not-openable window.

5. Discussion

This paper describes the development of a real-time IoT platform to monitor indoor air quality (CO_2, TVOC, Temperature, and Relative Humidity) and optimize it air exchanges. The proposed system is primary conceived to be used in residential spaces and in small offices to guarantee IAQ levels. Even if the proposed control methodology can be adapted to existing mechanical ventilation systems, by defining the minimal amount of required air exchanges for IAQ in accordance with internal monitored values, comfort thresholds, and external monitored conditions, its main application is in those spaces where no mechanical ventilation systems are installed. The low cost of this solution and the ease with which it can be integrated into buildings, e.g., not needing distribution channels, makes the system adaptable to existing and new buildings. Unlike other systems, the proposed one is a fully IoT solution in which each unit (e.g., external monitoring, internal monitoring and fan activation) is connected to the cloud infrastructure.

The modularity of this system is, in fact, very important, since it allows us to include more units and to be open to future improvements, thus providing adaptability and failure tolerance to the system. In addition, the IoT infrastructure can easily be adapted to be used in controlling current commercial detached mechanical ventilation systems. Furthermore, the proposed solution adopts the IAQI to indicate pollutant levels and by developing a mobile application, it raises user awareness of indoor air conditions, supplying a user-friendly and graphic interface, which makes the user a protagonist and guarantees user wellness. Early results from experimental tests show that the system can handle indoor air quality and keep it at good levels, by controlling indoor pollutants.

Future developments of this system are planned, including the possibility to implement machine-learning approaches to make the control strategy more effective and to forecast outdoor air quality evolution to adjust the ventilation schedule and prevent critical environmental conditions. Another important aspect to upgrade this system in future is its potential integration with existing models that define the passive moisture sorption/desorption potential of materials, by assuming, for example, the approaches to moisture buffering effects on IAQ proposed in Ref. [45–47]. In addition, the system has already been conceived to define different priorities for each pollutant in order to define personalized activation plans for each connected ventilation unit (e.g., fan). Specific analyses on this topic will be conducted in future. For example, target CO_2 values may be differentiated in accordance with indoor activities, such as between an adult's bedroom and a child's one. Moreover, the possibility to control each pollutant threshold independently will also be tested in real operational conditions. For example, CO_2 may be controlled to reach a good IAQI level, while $RH\%$ maintained at a moderate level. This can be done by acting on the settings of the mobile user interface—see Figure 5. It should be noted that, at present, the system has not been implemented to consider the IAQI temperature index when controlling fan activation, while in the future, additional considerations about thermal comfort thresholds and ventilative cooling modes will be included in order to give users more customizable options when setting the system. As regards cooling, the system will be implemented to guarantee a proper fan-driven ventilative cooling action. In this case, the system will be activated when outdoor temperatures are lower than those indoor in free-running buildings, or below the comfort threshold in conditioned spaces, see for example [27]. As regards heating, the system may be implemented in future to define a balance-control algorithm. This algorithm will consider ventilation requirements for acceptable pollutant levels, and heat losses through ventilation. With respect to fixed standardized air ventilation rates, the proposed system allows us to activate the fan in winter only to move the monitored-based airflow by supporting a smart ventilation approach which prevents over-ventilation. Nevertheless, an extended analysis is planned at a further developmental stage, which will relate IAQI index requirements and ventilation energy losses considering internal-external differences to suggest optimal balance points.

Finally, separate sensors included in the indoor/outdoor units should be integrated to a single circuit board to achieve high integration and guarantee the application of the Plug and Play style. The code has already been prepared for this further step, as it is the list of connected devices read by the device connector using the service catalog. In this way, the system is already ready to work on different pollutant sources in each room, e.g., by controlling the IAQI indexes of $RH\%$ in a bathroom, or CO_2 and TVOC in a living room. The elaborated IoT infrastructure is primed to be able to take into consideration additional pollutants. At present, the solution system has reached a TRL of 3 to 4, while higher levels of TRL are expected to be reached during further developmental stages. In particular, it is planned to test the system in real environments, with special regard to school rooms, including a comparison with a room which is not equipped with the IoT system, both in terms of heat loss due to ventilation and the ability to maintain the room in the best IAQI class. These further tests, in fact, will include validations in cold winter conditions. It must be remembered, however, that, this second developmental phase is subject to devoted funding being available to raise the system from a TRL 3–4 to a TRL 6.

6. Conclusions

The paper presents a new IoT system composed of different units (slaves) which are connected with the cloud and that communicate with a building control unit (master). The system focuses on both monitoring IAQ levels and guaranteeing IAQ comfort levels in building spaces at a low cost. The approach is fully scalable, and is able to connect, for example, different indoor microcontroller units to monitor the IAQI levels and activate independently different connected ventilation units.

The main outcomes are listed here below:

- The definition of a low-cost IoT system to monitor and guarantee optimal IAQ levels in building spaces without the need to install large mechanical ventilation systems;
- The proposed system is able to balance indoor and outdoor IAQ levels to define the proper airflow rate to reach the desired indoor IAQI index value;
- The system in the meantime, is able to process different pollutant sources by monitoring different sensors, and consequently set the minimal air exchange rate to guarantee optimal IAQI levels for all pollutants (Section 4.3). This approach is of course scaled in order to consider a different set of pollutants or to include priority lists, by modifying the desired levels of each pollutant independently as confirmed by early tests on single pollutants (Sections 4.1 and 4.2);
- The adoption of a full IoT infrastructure connecting each unit to the cloud to allow for different control levels and define a fully-scalable platform which is able to connect and control different independent units;
- The system, being based on monitored levels, is able to adapt to specific emission rates, to different environmental conditions, by adjusting the fan speed (PWM);
- The definition of simple user interface by the development of a mobile app, which is able to inform users and permit setting control. Two user interfaces were developed for non-expert users and for expert ones.

Current TRL is 3 to 4 though further developments are planned to implement the proposed solutions in order to develop it till commercialization.

Author Contributions: Conceptualization, G.C.; Data curation, S.C., M.G., M.I. and S.L.; Supervision, G.C.; Writing—original draft, G.C., S.C., M.G., M.I. and S.L.; Writing—review and editing, G.C.; Funding acquisition, G.C.

Funding: This research was co-funded by the 59_ATEN_RSG16CHG.

Acknowledgments: The proposed approach was tested during the Course IP, Master Degree in ICT for Smart Society, Politecnico di Torino, Italy, A.Y. 2018-19, with the support of the LASTIN laboratory, Microclimate section (GC). F. Dovis is thankfully acknowledge for precious suggestions together with C. Ward for language checking.

Conflicts of Interest: The authors declare no conflict of interest. The funders had no role in the design of the study; in the collection, analyses, or interpretation of data; in the writing of the manuscript, or in the decision to publish the results.

References

1. EPA. Introduction to Indoor Air Quality. Available online: https://www.epa.gov/indoor-air-quality-iaq/introduction-indoor-air-quality (accessed on 25 July 2019).
2. EPA. *Building Air Quality. Action Plan*; EPA 402-K-98-001; DHHS (NIOSH) No. 98-123m; EPA-NIOSH: Cincinnati, OH, USA, 1998. Available online: https://www.epa.gov/sites/production/files/2014-08/documents/baqactionplan.pdf (accessed on 25 July 2019).
3. Grün, G.; Urlaub, S. *Towards an Identification of European Indoor Environments' Impact on Health and Performance—Homes and Schools*; Fraunhofer-Institut für Bauphysik IBP: Stuttgart, Germany, 2014.
4. WHO Europe. *WHO Guidelines for Indoor Air Quality: Dampness and Mould*; World Health Organization, Regional Office for Europe: Copenhagen, Denmark, 2009; ISBN 7989289041683.
5. Prüss-Ustün, A.; Wolf, J.; Corvalán, C.; Bos, R.; Neira, M. *Preventing Diseases through Healthy Environments: A Global Assessment of the Burden of Disease from Environmental Risks*; World Health Organization: Geneva, Switzerland, 2016; ISBN 978 92 4 156519 6.
6. Passive House Institute. Available online: https://passivehouse.com/index.html (accessed on 8 July 2019).
7. Consoli, A.; Costanzo, V.; Evola, G.; Marletta, L. Refurbishing an existing apartment block in Mediterranean climate: Towards the Passivhaus standard. *Energy Procedia* **2017**, *111*, 397–406. [CrossRef]
8. Gross, A.; Mocho, P.; Plaisance, H.; Cantau, C.; Kinadjian, N.; Yrieix, C.; Desauziers, V. Assessment of VOCs material/air exchanges of building products using the DOSEC®-SPME method. *Energy Procedia* **2017**, *122*, 367–372. [CrossRef]
9. WHO Europe. *Air Quality Guidelines. Global Update 2005—Particulate Matter, Ozone, Nitrogen Dioxide and Sulfur Dioxide*; World Health Organization, Regional Office for Europe: Copenhagen, Denmark, 2006.
10. WHO. *Indoor Air Quality Guidelines: Household Fuel Combustion*; World Health Organization: Geneva, Switzerland, 2014; ISBN 9789241548878.
11. WHO Europe. *WHO Guidelines for Indoor Air Quality: Selected Pollutants*; World Health Organization, Regional Office for Europe: Copenhagen, Denmark, 2010; ISBN 9789289002134.
12. Allen, J.G.; MacNaughton, P.; Satish, U.; Santanam, S.; Vallarino, J.; Spengler, J.D. Associations of Cognitive Function Scores with Carbon Dioxide, Ventilation, and Volatile Organic Compound Exposures in Office Workers: A Controlled Exposure Study of Green and Conventional Office Environments. *Environ. Health Perspect.* **2016**, *124*, 805–812. [CrossRef] [PubMed]
13. Arundel, A.V.; Sterling, E.M.; Biggin, J.H.; Sterling, T.D. Indirect health effects of relative humidity in indoor environments. *Environ. Health Perspect.* **1986**, *65*, 351–361. [CrossRef] [PubMed]
14. Chiesa, G. Data, BigData and smart cities. Considerations and case study on environmental monitoring. *Techne* **2014**, *8*, 81–89. [CrossRef]
15. Chiesa, G. Explicit-digital design practice and possible areas of implication. *Techne* **2017**, *13*, 236–242. [CrossRef]
16. Netatmo Weathermap. Available online: https://weathermap.netatmo.com/ (accessed on 26 July 2019).
17. Libelium. *Smart Gases Pro: Technical Guide*; Waspmote. Document, Version 7.6; Libelium Comunicaciones Distribuidas S.L.: Zaragoza, Spain, 2019; 119p, Available online: http://www.libelium.com/downloads/documentation/gases_sensor_board_pro.pdf (accessed on 17 October 2019).
18. Libelium. *Smart City Pro: Technical Guide*; Waspmote. Document Version 7.6; Libelium Comunicaciones Distribuidas S.L.: Zaragoza, Spain, 2019; 96p, Available online: http://www.libelium.com/downloads/documentation/smart_cities_pro_sensor_board.pdf (accessed on 16 October 2019).
19. Mohieddine Benammar, M.; Abdaoui, A.; Sabbir, H.M.; Ahmad, S.H.M.; Touati, F.; Kadri, A. A modular IoT platform for real-time indoor air quality monitoring. *Sensors* **2018**, *18*, 581. [CrossRef]
20. Sun, Z.; Wang, S.; Ma, Z. In-situ implementation and validation of a CO_2-based adaptive demand-controlled ventilation strategy in a multi-zone office building. *Build. Environ.* **2011**, *46*, 124–133. [CrossRef]
21. Shin, M.S.; Rhee, K.N.; Lee, E.T.; Jung, G.J. Performance evaluation of CO_2-based ventilation control to reduce CO_2 concentration and condensation risk in residential buildings. *Build. Environ.* **2018**, *142*, 451–463. [CrossRef]

22. Zucker, G.; Sporr, A.; Garrido-Marijuan, A.; Ferhatbegovic, T.; Hofmann, R. A ventilation system controller based on pressure-drop and CO_2 models. *Energy Build.* **2017**, *155*, 378–389. [CrossRef]

23. Carrilho, J.D.; Mateus, M.; Batterman, S.; Gameiro da Silva, M. Air exchange rates from atmospheric CO_2 daily cycle. *Energy Build* **2015**, *92*, 188–194. [CrossRef] [PubMed]

24. Costanzo, V.; Yao, R.; Xu, T.; Xiong, J.; Zhang, Q.; Li, B. Natural ventilation potential for residential buildings in a densely built-up and highly polluted environment. A case study. *Renew. Energy* **2019**, *138*, 340–353. [CrossRef]

25. RENSON Mechanical Ventilation. Available online: https://www.renson.eu/gd-gb/products/ventilation/mechanical-ventilation (accessed on 26 July 2019).

26. Holzer, P.; Psomas, T. (Eds.) *Ventilative Cooling Sourcebook*; IEA EBC Annex 62; Aalborg University: Aalborg, Danmark, 2018; ISBN 87-91606-37-3.

27. Heiselberg, P. (Ed.) *Ventilative Cooling Design Guide*; IEA EBC Annex 62; Aalborg University: Aalborg, Danmark, 2018; ISBN 87-91606-38-1.

28. CherryPy. Available online: https://cherrypy.org/ (accessed on 26 July 2019).

29. MQTT. Available online: http://mqtt.org/ (accessed on 26 July 2019).

30. Eclipse Mosquitto. An Open Source MQTT Broker. Available online: https://mosquitto.org/ (accessed on 26 July 2019).

31. MongoDB. Available online: https://www.mongodb.com/ (accessed on 26 July 2019).

32. Bosch. BME680 Datasheet. Available online: https://cdn-shop.adafruit.com/product-files/3660/BME680.pdf (accessed on 16 September 2019).

33. U.S. Environmental Protection Agency (USEPA). *Guidelines for Reporting of Daily Air Quality- Air Quality Index (AQI)*; Series EPA-454/B-06-001; U.S. Environmental Protection Agency (USEPA): Research Trangle Park, NC, USA, 2006.

34. Saad, S.M.; Shakaff, A.Y.M.; Saad, A.R.M.; Yusof, A.M.; Andrew, A.M.; Zakaria, A.; Adom, H. Development of indoor environmental index: Air quality index and thermal comfort index. *AIP Conf. Proc.* **2017**, *1808*, 020043. [CrossRef]

35. Nigam, S.; Rao, B.P.S.; Kumar, N.; Mhaisalkar, V.A. Air Quality Index—A Comparative Study for Assessing the Status of Air Quality. *Res. J. Eng. Technol.* **2015**, *6*, 267–274. [CrossRef]

36. Applied Sensor. *Ventilation Energy Savings While Maintaining Premium Air Quality*; Applied Sensor: Reutlingen, Germany, 2013.

37. Environmental Bureau, Environmental Protection Department. Chapter 4: Government initiatives to improve indoor air quality. In *Director of Audit*; Report No. 56; Audit Commission: Hong Kong, China, 2011.

38. Department of Occupational Safety and Health. *Industry Code of Practice on Indoor Air Quality 2010*; Ministry of Human Resources: Putrajaya, Malaysia, 2010; ISBN 983201471-3. Available online: http://www.dosh.gov.my/index.php/legislation/codes-of-practice/chemical-management (accessed on 25 July 2019).

39. ASHRAE. Ref. CO2 requirements in 6.1.3 Ventilation Requirements. In *Interpretations for Standard 62-1989*; ASHRAE: Atlanta, GA, USA, 1995; Interpretation 62-1989-20.

40. Department for Education. *Guidelines on Ventilation, Thermal Comfort and Indoor Air Quality in Schools*; Department for Education: London, UK, 2016.

41. Nicol, F.; Humphreys, M.; Roaf, S. *Adaptive Thermal Comfort: Principles and Practice*; Earthscan-Routledge: London, UK, 2012. [CrossRef]

42. Chiesa, G.; Grosso, M.; Bogni, A.; Garavaglia, G. Passive downdraught evaporative cooling system integration in existing residential building typologies: A case study. *Energy Procedia* **2017**, *111*, 599–608. [CrossRef]

43. Cetiat. *Tables de l'Air Humide*; CETIAT: Villeurbanne, France, 1976.

44. Patel, M.A.; Patel, A.R.; Vyas, D.R.; Patel, K.M. Use of PWM techniques for power quality improvement. *Int. J. Recent Trends Eng.* **2009**, *1*, 99–102.

45. Di Giuseppe, E.; D'Orazio, M. Moisture buffering active devices for indoor humidity control: Preliminary experimental evaluations. *Energy Procedia* **2014**, *64*, 42–51. [CrossRef]

46. Balocco, C.; Petrone, G. Heat and moisture transfer investigation of surface building materials. *Math. Model. Eng. Probl.* **2018**, *5*, 146–152. [CrossRef]
47. Salonvaara, M.; Ojanen, T.; Holm, A.; Künzel, H.M.; Karagiozis, A.K. Moisture buffering effects on indoor air quality. Experimental and Simulation Results. In Proceedings of the Buildings IX: Thermal Performance of Exterior Envelopes of Whole Buildings International Conference, ASHRAE, Atlanta, GA, USA, December 2004.

MDPI

St. Alban-Anlage 66

4052 Basel

Switzerland

Tel. +41 61 683 77 34

Fax +41 61 302 89 18

www.mdpi.com

Sustainability Editorial Office

E-mail: sustainability@mdpi.com

www.mdpi.com/journal/sustainability